THE CONCEPT OF TIME
IN
ANCIENT INDIA

THE CONCEPT OF TIME
IN
ANCIENT INDIA

By

Dr. RALLAPALLI VENKATESWARA RAO

M.A. (Ind.Cult), M.A. (Jyotisha), M.Phil., Ph.D.,
P.G. Diploma in I.R. & P.M., Museology, Jyotisha

BHARATIYA KALA PRAKASHAN

DELHI (INDIA)

First Published : 2004

Revised Edition : 2024

Published by C.P. Gautam *for*
BHARATIYA KALA PRAKASHAN
3480/1, IInd Floor, Narang Colony, Tri Nagar,
Delhi - 110035, Mob.: 9811642997, 8595139625
E-mail : info@bkpbooks.com; sales.bkpbooks@gmail.com
Website: www.bkpbooks.com

ISBN : 978-81-8090-032-7

Laser Type Setting by
A-ONE GRAPHICS
X-4, Street No. 2, Brahmapuri, Delhi-110053

Printed at
Excel Prints, New Delhi-28. Ph. 011-25776555,

Dedicated

to

My Beloved Parents

CONTENTS

FOREWORD

Time is one of those things which are beyond the comprehension of the human mind and therefore defy any definition in spite of the fact that different views are expressed by different thinkers on this subject. The very fact that so many theories have developed around the concept of mind is a proof to maintain that Time is Anirvacanīya (beyond definition).

For the Vaiśēṣikās Time is an eternal, all pervasive substance. For others it is a standard of measurement for actions : an action of known dimensions used as a standard for measuring other action. For some, it is nothing more than causality. For others still, it is mere construction of the mind or identical with the three guṇas.

For Bhartṛhari, the great grammarian, time is the creative power (Kartṛ-Śakti) of Brahman. It is the Svatantrya Śakti, power of complete freedom of Brahman, as a result of which the whole phenomenal world is presented to us in temporal sequence. The sequence is super imposed on the phenomena. As the sequence originates from the self, the latter is figuratively called Time. He is the ground on which the effect of time, namely sequence can be seen.

As there is no difference between power and that which wields it. Time is really Brahman. As the creative power, Time is responsible for the birth, continuity and destruction of every thing in the Cosmos. Every thing has thrown special cause; yet all those special causes depend upon Time.

Time is compared to the stage manager (Sūtradhāra) of a puppet show who pulls the strings and makes the puppets dance according to his wish. The whole Cosmos is a collection of puppets and Time controls their performances.

Time has two functions-Abyanujña (permission) and pratibandha (prevention). Somethings appear at a particular time while others do not appear at that time. If a thing is produced at a particular time, it is because time allows the powers of its cause to be effective at that time. This function of time is called Abyanujña, If something does not appear at any time, it is because Time prevents its appearance through its second function, Pratibandha.

Tamasya Lōkamantasya Sūtradhāraṃ Pracakṣatē/
Pratibandhābhyanujñābhyāṃ Tēna Viśvaṃ Vibhajyēta//

—Vākyapadīya III

But for these two functions of Time there would be confusion in the Universe. Thus Time is almost identical with Brahman, as it is its power, according to the grammarians.

In the Trika siddhānta of Kāśmīr, Time is not accepted as having an independent existence. It is Śiva that causes sequence in the time by the appearance of various actions.

kriyāvaicitr[illegible] ɪnirbhāsāt Kālakramamapīśvaraḥ/
sarvatrābhāsabhēdōpi Bhavēt Kālakramākaraḥ//

—Īśvaraprabhu Vijñānakārikā (2.1.5, 6)

According to the Viśiṣṭādvaita, Time is a substance eternal and all pervading and it is without the three guṇas. This is responsible for all kinds of usages (Vyavahārās) like 'before' 'after' 'simultaneously', 'long ago', minutes, days, weeks etc. This is helpful to the Lord Viṣnu in regulating all the affairs connected with the phenomenal world and He depends upon it in all his actions connected with it.

For Advaitas there is no such thing as Time, because it is neither perceived by the senses nor inferred on the strength of any Hētu. It is not correct to say that Time is responsible for the usages such as 'earlier' and 'anterior' etc. In fact, it is not Time that is responsible for such usages, but such usages that are responsible to imagine that there is such a thing called Time. All the Vyavahārās that are attributed to Time, the all pervading eternal Ātman (consciousness) is the ground. Therefore, Time is only a product of imagination for the Advaitin. But the grammarians or the followers of Trika Siddhānta or the Advaitins are not against accepting Time at the Vyavahārika lever with all its divisions. There are some of the important views on Time of Indian philosophers.

A thorough study of all these important theories has been made by Dr. Rallapalli Venkateswara Rao in his valuable work on Time. Starting his investigation about the nature of Time from the oldest Indian texts like the Vedās, the author proceeded to serutimise Mahābhārata, Bhagavadgīta, part of Bhārata, Purāṇas and other works explained the different views on Time contained in all those ancient works. He has examined the views of Jains and Buddhists also whose contribution on subjects like this is of great significance. He has ultimately concentrated on what is found in the Purāṇas and in Jyōtiśśātra and in doing so he used closed acquaintance with the Indian Astrology to the fullest possible extent.

This work abounds with quotations from various ancient works starting with the Vēdās which are highly informative, though some of them apparently have little bearing on the topic under discussion.

This is an important work on the Concept of Time which I hope will be welcome by both philosophers and astrologers and use in their further investigations on this subject.

Place : Hyderabad
Date : 19th September 2003

Dr. P.S. Ramachandrudu
Satadhika Grantha Nirmata
Mahamahopadhyaya
M.A., Ph.D. (Skt), M.A.(English), M.A.(Hindi),
Recipient of President's Certificate of Honour
"Vishwabharati Award" and two "Sahitya Academi Awards"
Professor of Sanskrit, Osmania University (Retd.)

MESSAGE

Dr. Rallapalli Venkateswara Rao of the Potti Sriramulu Telugu University has written an interesting and erudite treatise on the various interpretations of Time in Indian thought over the millennia. The topic is very important, with its antiquity and contrast with concepts in other cultures. It is a constant source of scholastic interaction and no doubt promises to be so in future also.

Obviously, there can be no single unanimous of definitive opinion on the subject, ever. The debate and discourse will continue for a long time.

However, the present thesis is written after considerable effort and study for which Sri Venkateswara Rao deserves compliments.

I would like to congratulate Dr. Venkateswara Rao for his well-researched thesis and hope that his book will evoke renewed interest in intellectual circles and others, since it brings out the splendid effort that was made in India since times immemorial.

Hyderabad
7th August 2003

P.V. Narasimha Rao
Former Prime Minister of India

ACKNOWLEDGEMENTS

I owe a debt of gratitude to several individuals and institutions for their help and encouragement in the completion of this Thesis. In the first place, I ever remain deeply indebted to my mother Mrs. R. Jayalakshmi and father late Sri R. Vallabha Rao for their constant encouragement given to me.

I take this opportunity to express my deep sense of gratitude to Dr. M. Radha Krishna Sarma, Professor and formerly Head of the Department of Ancient Indian History, Culture and Archaeology, Osmania University, under whose supervision, I have completed the thesis. In spite of many pressing engagements, he spared his precious time and gave scholarly advice at every stage in the preparation of the thesis. His abiding and earnest interest in my work and concern for my well being were an invaluable source of strength and inspiration to me.

I express my gratitude to Prof. Dr. (Mrs.) Dhareswari, Department of Ancient Indian History, Culture and Archaeology, Osmania University for her kind suggestion and encouragement given to me.

I am highly grateful to the Osmania University which not only permitted me to undertake this Research but also provided the financial assistance required through a scholarship for Ph.D. Course.

I express my deep sense of gratitude to the Heads of various Libraries for allowing me to have access to them.

My thanks are due to the Librarian and Staff especially, I Satyanarayana, Asst. Librarian, Srinivasa Rao, Book Catalogue Section and Venkatiah, Book-Keeper of O.U. Library, who gave me unstinted help and cooperation in consulting research material.

I will be failing in my duty if I do not acknowledge the help of Sri M. Niranjan Shah, Audit Officer, Local Fund Audit Department, Hyderabad, who offered valuable suggestions in preparing this Thesis.

I am also grateful to Madhavi, Bala, Sundari, Mallik for their help in the collection of material and to Dr. Srinivas Sarma, Sanskrit Academy for his help in the translation of various Sanskrit verses.

I record my gratitude to Sri K.V.N. Mani Gopal, Stenographer, Canteen Stores Department, Ministry of Defence, Visakhapatnam, for his help in typing the Thesis. I must express hearty thanks to Sri C.P. Gautam, Bharatiya Kala Prakashan and his staff members for their kind cooperation in bringing our this publication with diligent execution.

26th Jan. 2003 **Dr. R. Venkateswara Rao**

Hyderabad.

INTRODUCTION

The concept of Time has always been associated with the human experience of the changing phenomena, so much so that Time is conceived as the single most coordinating factor of all experiences and their interpretations. This gives rise to a belief in its metaphysical reality; a reality which seems to be constitutive of our experiences and thinking. The source of the experience of time is basically the all-round change and periodicity in the eternal world. Thus the experience of the concomitant of time as a coordinative factor with all other experiences and thinking was the main reason of reckoning Time in various ways-namely, as Cosmic Principle, Universal Cause, God and as the Originator, Sustainer and Destroyer of the Universe. The first conscious reflections on the problem of Time, in India, are found in the Atharvavēda, which give the descriptions of Time in these terms.

Since the dawn of human thinking and reflections on experiences, the Indian mind has been engaged in the search for the meaning, aim and ideal of life. Its all out effort, motivated by the spiritual urge, was to go beyond the sensible phenomena in order to know their underlying principle. In this process different philosophical schools of thought came into existence. Because of the openness of their minds, the ancient Indian thinkers in general gave prominence to some or the other of the four traditional human values. These values are : Material or Economic Value *(Artha),* Value of Sensual

Pleasure or Enjoyment *(Kāma),* Morality or Moral Value *(Dharma)* and Spiritual or Esoterical value *(Mōkṣa)*.

In the course of the development of thought most thinkers and their followers, except Cārvāka undertook the task of pursuing the fourth value the summum bonum of human existence. So the pursuance and inculcation of the fourth value, i.e., *Mōkṣa* became the most sought-after aim in such a way the seekers in the due course rigorously developed different, and sometimes contradictory schools or systems of thought although they had a common goal. This situation arose because of their choice or preference for one set of basic elements to another, and their subsequent methodological approach, which gave rise to their different conceptual frame works, convictions or orientations. Thus, it was natural to have different metaphysical systems in India. The elements other than the basic ones of a system were set and interpreted according to the philosophical needs of its frame works. This is the reason why the concept of Time has resulted in different conceptions.

Despite all differences, the concept of Time was never considered in isolation, because it is impossible to conceive of Time independently of the allied concepts of being and becoming, change and causality and so on. This explains why we get different conceptions of Time in the *Vēdas. Upaniṣads, Purāṇas* and the various philosophical schools of *Sāṅkhya, Yōga, Nyāya, Vaiśēṣika,* Jainism and Buddhism and why Time is treated differently as real or unreal, subjective or objective, primary or secondary, physical or psychological, creator or created, interiorised or exteriorised, substantive or attributive, empirical or a priori and so on. Therefore, an understanding of the nature of Time and its pertinence to the life of man is essential to the approach of a fairly adequate philosophic vision of reality. Hence, the study on "The Concept of Time in Ancient India" has been undertaken.

The review of research and literature on the subject reveals that the dimensions of the topic are vast and deep. Though the efforts made so far provide documentation on different aspects, a discerning examination of the studies reveals that there were no specific studies on this topic.

Further, there were no specific studies on the chronology of the Indian sub-continent determining properly the number of Eras used in the relevant inscriptions and on fixing at least approximately the year or years of such reckoning or reckonings. Periodic review of the problem is therefore a desideratum for a correct evaluation of the History. This necessity seems to have been further enhanced by the archaeological material discovered. Unfortunately all this data as well as other information have not yet been fully utilized in one single work. So an attempt is made to fill this gap.

Objectives of Study

1. To study the Concept of Time in Ancient India in *Saṃhitas, Brāhmaṇas, Upaniṣads, Epics* and Purāṇas.
2. To study the Concept of Time in various philosophical schools, viz., *Nyāya, Vaiśēṣika, Sāṅkhya, Yōga, Jainism* and *Buddhism.*
3. To examine and assess the division of Time and its computation in Ancient India.
4. To study the Concept of Time from Archaeological evidences like Inscriptions.
5. To find out the significance of Time and its pertinence to the life of man, viz., times of worship, feast days and propitious time for a particular undertaking.

Hypotheses

The following hypotheses have been formed and tested :

1. The Cosmic significance and Metaphysical Truth of Time by Indian minds in search for the meaning, aim and ideal of life.
2. Time has been a priori explanatory principle of all changes, events, happenings, processes and subsistence of Being.
3. Time has been the single most coordinating factor of all human experiences and reflections.
4. Time played the vital role in all walks of life of human being and his existence in Ancient India.

Methods of Data Collection

This thesis is based upon the Historical method. The material is collected from mainly literary works and archaeological sources. The literary works include the Vēdic literature, *Brāhmaṇas, Upaniṣads, Epics* and *Purāṇas* and also the Contemporary Texts, Journals and the Periodicals. The archaeological sources include Inscriptions, Plates published in various works like E.I., J.A.H.R., A.S.I. Annual Reports, Indian Antiquary, Asiatic Researches etc.

Chapterization

The present study is divided into the following Chapters :

Introduction to the Subject, which explains the choice of the Research topic, Objectives of the study, Hypotheses, Methodology and Chapterization.

The First Chapter deals with the Concept of Time in the *Vēdic* and *Post-Vēdic* periods. The mention of Time in the *Vēdas,* Time as fundamental and very important in the process of evolution, Time as above everything else, even above God, and as the actual existence of the beings, cultic time etc., are taken in the Vēdic period. The transition from the cultic time of the *Vēdas* to the interiorized time of the *Upaniṣads,* comparison of time with Brahman and mention of Time as a Cosmic power different views about time in *Mahābhārata,* the *Purāṇic* conception of Time as divinity and their attempt to integrate it in their respective theologies, the eternal substantial concepts of *Nyāya, Vaiśēṣika* schools and the Sāṅkhya conception of Time as the present moment identical with the unit of change of the *Guṇas* etc., are taken in the Post-Vēdic period.

The Second Chapter is devoted to the Concept of Time in *Jainism* as a real substance. Attempt was made on the contributions of Jains in the domain of Cosmology and Time.

The Third Chapter deals with the Concept of Time in *Buddhism.* Buddhist conception of time and temporality and their consideration of time and causation as parts of our experience are taken into account.

The cyclic time which occurs frequently in Cosmological context such as Hindu theory of *Yugas, Manvantaras* and *Kalpas* and Linear Time in historical sources such as Genealogies, Biographies and Chronicles where time reckoning was recorded through generations, regnal years and eras are given in the Fourth Chapter.

The Fifth Chapter deals with the Pre-Kaliyuga and Kaliyuga Eras. The reckoning of the natural divisions of Time-Years, Months, Days by the motion of the Sun and the Moon, the four Yugas, i.e., Kṛta, Trēta, Dvāpara and Kali with number of years, the Eras like Saptarṣi-Kāl or the cycle of the Seven Ṛṣis, the Bārhaspatyamāna or 60 and 12 year cycles of Jupiter and the Kaliyuga or beginning of the Kali age and the Eras after Kali are taken into account.

The Sixth Chapter is devoted to the Significance of

Applications of Time. The auspicious time for making *Dāna* which is mentioned in the inscriptions by various donors, the mentioning of the Year, *Āyana,* Month, Fortnight, *Tithi* or Lunar Day, *Vāra* or Weekday, *Nakṣatra, Yōga, Karaṇa, Muhūrta, Grahaṇa, Saṅkrānti* and their importance in making *Dāna* in different *Dharma Śāstras* are taken into account. The epigraphical instances are also given in the form of tables.

At the end, summary of the entire study is given.

CHAPTER - I

CONCEPT OF TIME IN THE VEDIC AND POST-VEDIC PERIODS

The Vēdas are the oldest authority on Indian philosophy. The word *'Vēda'* is derived from the Sanskrit word *'vid'* (to know) and thus the word Vēda means the knowledge. The Hindus consider the Vēdas to be revealed books and give them the titles of *Apauruṣēya* (not made by man) and *Nitya* (eternal). It is contented that the *Ṛṣis* wrote them under inspiration from God. However, it is pointed out that the Vēdic hymns were composed by ancient *Ṛṣis* or seers of great antiquity. They were handed over from generation to generation. Thus they came to be known as the *'Śṛti'*. The Vēdas according to them enshrine the eternal and ultimate truths.

There are four Vēdas viz., Ṛg, Yaj, Sāma and Atharva. Every Vēda comprises three parts : Saṃhitā, Brāhmaṇa and Upaniṣad. The Saṃhitās are the collection of mantras. The *Saṃhitās* and *Brāhmaṇas* deal with *Karmakāṇḍa*. Philosophical thoughts are enshrined in the *Upaniṣads* and *Āraṇyakas*.

VEDIC PERIOD

There is mention of Time in the Vēdas and important characteristic of this period is to treat Time to be fundamental and very important in the process of evolution. Sometimes Time is made to occupy a place above everything else, even above God. The existence of Universe and the history of man and the cosmos come under the sway of two superior forces-*Kāla* (time) and *Karma* (the act).

In the earliest experience of Vēdic period, time was perceived as actual existence of the beings, we describe as temporal. There is no such thing as empty time. Time is an abstraction, which does not exist. What does exist is the (chronological), flux of Beings; and it is the process, which makes Sacrifice possible. Hence, Time is experienced as the fruit of Ritual action.

Time is born with sacrifice, and it is by sacrifice that is once again destroyed. This concept is at the root of the intimate relationship between worship and Time, and provides us with a key to understanding of the central place of sacrifice and man's participation in the unfolding of Time. In this sense Time is something which man makes in close collaboration with God's time, i.e., the continuing existence of beings, is a theandric product, in the Vēdas, in the Saṃhita period, we find several words to designate Time, for example, *Āyus,* life time, life span[1] or *Ṛtu* the right time for sacrifice, the season.[2] Abstract time is of no interest to the Ṛṣis of the Vēdas for whom there is no continuity of time apart from Ritual activity or the act of God (for instance Indra).

"This time.....has no reality that is to say efficiency, except in the moments in which divine or sacred acts are concerted.....In this succession of acts linking moments it would be vain to look for a given continuity : Continuity is no more than the fruit of the constructive activity which recommences day after day.[3]

In the Vēdas the unit of time is the day, which is the centre of all experience of time.[4] Dawn and twilight are "Junctions", the most critical moments in the whole day. It is from day to day (divē divē)[5] and by dint of the daily sacrifice, the *Agnihōtra,* that time endures and existence continues.

Later as the sacrifice became increasingly elaborate and the building of the fire altar in the *Brāhmaṇas* stretched out over a year, it was the year, which became the larger unit of time. The sacrifice remained the foundation of the temporal structure, each brick of the altar corresponding to one day in the year.

A fundamental reflection on time, if it does not beg the question from the very beginning by assuming that we already know that time is, has to take into account what Man has understood time to be. Because neither man nor time is objectionable entities, Man's self-understanding belongs to what Man is, and similarly Man's opinions on time can't be excluded from a fundamental research on the nature of time. In other words, we have to draw from the global human experience if we want to say something about the Universal deity... speaking with the Vēdas...that we call by the name of Time. The purpose is to show the intimate connection between the human experience of time and the equally human practice of sacrifice.

We may describe the quintessence of the Vēdic tradition by mentioning a few traits. It is by the Sacrifice that not only Man but also God acquires immortality. By means of the Sacrifice, man reaches heaven, freedom, happiness, purification of senses, superiority. It is the sacrifice that helps Man overcome his earthly condition. By performing the sacrifice, Man reaches the navel of the world.[6] Authentic existence is reached only in and through sacrifice. By the sacrificial act, Man reaches the shores of the other world and is saved from the grip of Time.

In point of fact, the Brāhmaṇic literature will stress that Man is capable of attaining the full span of his life, for Man's immortality is a full life span."

The subsequent Jñānic spirituality will go a step further and proclaim that through the spiritual sacrifice Man is saved from time altogether, that is released. To sacrifice is the highest action because in it Man collaborates with the world, and the Gods to sustain the entire Universe. In a word, all these and many other utterances make sense only if we discover in sacrifice the ultimate texture of reality. Namely, that act by which the Universe has come into being[7] and by which it is maintained.

One trait emerges from all this : sacrifice is every act by which Man partakes in the power of a supratemporal agency. He reaches beyond the immediate present into the recesses of the past, sometimes even to the beginning of Time, and also into the reservoir of the future, in some cases even to the end of Time. The sacrificial act may also serve to acquire this or that earthly desire, but ultimately it differs from every other action because it stretches beyond the ordinary time-span and, rescuing Man from his servitude to time, truly liberates him.

It was the *Puruṣa,* the Cosmic Man of the *Ṛgvēda*[8] who was originally immolated in order that the world might exist : the world exists only by virtue of this primordial sacrificial act. So, in the Vēdas this activity is frequently compared to that weaving,[9] the well being made up of the day and the night[10] and ritual moments.[11]

Thus, in this intuitive Vēdic view of time there is, first, the idea of a relationship between time and the act of worship (*Karma* in the intrinsic sense of the word) which is so close that one does not exist without the other; and secondly, Vēdic man-unlike man in subsequent periods-aspires either to long life, or to a certain kind of continuity which does not seem to be guaranteed by the cosmological events.[12]

The second fundamental intuitive notion of time, which is moreover akin to the time as the fruit of ritual action, goes so far as to consider time as a cosmic power.

The two famous Kāla hymns *(Kāla-Sūktas)* of the Atharvavēda-XIX 53 & 54,[13] in which we find graphically enshrined the idea of time as the supreme ruler of the Universe.[14]

The exaltation of great Time in these two hymns of *Atharvavēda* is the earliest expression of this vision of Time as 'the Creator of the Creator', Prajāpati, who is Brahman (the ultimate principle of the Universe) itself :—

1. Time draws (the chariot) like a horse with seven reins,
   a thousand-eyed, fruitful-loined, immune to age.
   Astride it are poets who understand inspired songs.
   Its wheels are everything that exists.

2. Thus time draws seven wheels,
   It has seven hubs, its axis is (called) non-death;
   On the hitherside of all these existences
   It advances, first among the Gods.

3. A full vessel has been placed above Time.
   We see (Time) even though it is
   In many places (at once).
   Opposite all these existences
   Time (is also seated), they say, in the highest firmament.

4. In oneness Time bore these existences,
   In oneness it encompassed them around.
   Time the father became Time their son.
   No glory higher than his.

5. Time engendered Heaven above,
   Time also (engendered) the Earths we see.

Set in motion by Time, things which were
And shall be are assigned their place.

6. Time created the earth;
In time burns the sun;
In Time (yes), Time, thc eye sees far off
All existences.

7. In Time is consciousness; in Time,
Breath; in Time is concentrated the name.
As Time unfolds
All creatures rejoice in it.

8. In Time is (sacred) Fervour, in Time (yes) in Time
is concentrated the all-powerful Brahman.
Time was the father of Prajāpati.[15]

9. By him this (Universe was urged forth, by him it was
Begotten, and upon him this Universe was founded.
Time Truly, having become the Brahma (spiritual exaltation),
Supports Paramēṣṭin (the highest Lord)[16]

"…..having by means of the Brahman (spiritual exaltation) conquered all the worlds, Time, the highest God, foresooth, hastens onward".[17]

All reality depends on Time, and even sacrifice, which elsewhere in the Vēdas is considered as the Supreme force, is likewise subordinated to Time. It is important to note the relationship, spoken of in practically all texts on time, between Absolute Time and Empirical Time, Creator and the Creative, father and son[18] cause and effect.

Here space is supported by, and extended in time.[19] Even inner realities—consciousness and breath—are under the sway of time.[20] A Universal dynamism sets everything in motion. To put it succinctly, *"Kāla"* has the supreme divinity which is subject neither to the personified Creator *(Prajāpati)* nor to the impersonal universal powers of the sacrifice or of the Brahman. "After conquering all worlds by the word, Time advances, the Supreme God".[21]

The *Ṛgvēda* also implies the Term *"Kāla"* like the word *Ṛtu* not in the general sense of Time but in that of "certain moment", "Right moment". In the latter meaning *Kāla* was often used in connection with the sacrificial ritual and although the etymology of the word has not yet been cleared up in a completely satisfactory manner.[22] The sense "right moment for a purpose" seems to be the original one, as Coldenberg has been able to show.[23] On the other hand, it has been suggested that *Kāla* may have meant at first something like "driver" and that this figure was also called "Special God" in the sense accepted by usener.[24] In a similar manner Guntert has tried to explain the character of *Savitar*.[25] But it is in the *Atharvavēda* that *Kāla* is mentioned for the first time as a God and it presents *Kāla* as the highest principle as the creator of the *Prajāpati* and the Universe. On the other hand, *Atharvavēda* says that *Rōhita,* a solar deity is *Kāla* and *Prajāpati*.[26] Whereas in the *Ṛgvēda, Rōhita* designs a red horse, and the conception of the Sun as a horse is well known. *Atharvavēda* calls *Kāla* a horse with seven reins and a thousand eyes. So, Atharvavēda retains notions of Time as the upper most God.

POST-VEDIC PERIOD

Upaniṣads

In the history of Indian philosophy the *Upaniṣads* represent the age when Indian philosophy, originating in *Vēdas* and passing through the narrow and hard ground of the *Brāhmaṇas* divided into many under-currents finding suitable ground in the Upaniṣads so much so that it became difficult to imagine its small origin by seeing its present vast form.

The term *Upaniṣad* literally implies "sitting near". It means sitting down near the teacher to receive instructions. Gradually the word *Upaniṣad* itself came to mean what was received from the teacher, a sort of secret doctrine, *"Rahasya"*, concerning the relation between the Creator and the created individuals. This generally held that 108 *Upaniṣads* were written by various saints and sages. However, the ancient *Upaniṣads* are the *Bṛhadāraṇyaka Upaniṣad, Chāndōgya Upaniṣad, Taittirīya Upaniṣad, Aitarēya Upaniṣad, Kēna Upaniṣad, Kauṣītaki Upaniṣad.* The *Upaniṣads* do not present a consistent system of philosophy but contain the opinions and lessons of many men who knew philosophy and religion. The fundamental doctrine which pervades all the genuine *Upaniṣads* can be summed up in the following sentence :

"The Universe is the *Brahman,* but the *Brahman* is the *Ātman"*. In other words, "the world is God, and God is My soul". The two conceptions of *Brahman* and *Ātman* are united in the philosophy of *Upaniṣads*.

The *Vēdas* sought the continuity of Time through the sacrificial act, but the *Upaniṣads* began to question the permanence of this act and this continuity.[27] Immortality, the whole concern of the Upaniṣadic sages[28] was no longer ensured by the performance of the rite. Continuity was no longer to be found externally, in the ritual or the Cosmos, but internally within the man or more exactly within the self, the *Ātman*. And yet, Cosmic connotations are not lacking in this new vision.

One of the first factors to be described in the search for

this continuity is the breath of life, *Prāṇa.*[29] *Prāṇa* is in the first place, the life principle and the individual aspect *(Ādhyātma)* of the unabating, Omni-present cosmic wind *(Vāyu);*[30] secondly, Prāṇa is not merely physiological breath but the Rhythm of respiration which also becomes a spiritual exercise *(Vrata)* to overcome death.[31] This is the beginning of Yōga exercises to control breathing *(Prāṇāyāma)*. If even the Sun is set to rise and fall in the Prāṇa, one begins to realise the Cosmological importance of the breath. Later on *Prāṇa* is identified with immortality *(Amṛta)*[32] and *Brahman* itself. The important point is that respiration corresponds to an internal time, and it is the mastering of this internal Rhythm especially in *Yōga,* that leads to the transcending of Time both externally and internally.

The transition from the *Cultic Time* of the Vēdas to the *Interiorised Time* of the Upaniṣads occurs evidently at the point where respiration interpreted as a sacrifice, takes the place of the sacrifice of the fire *(Agnihōtra) (Prāṇāgnihōtra Upaniṣad)*.

In addition, the *Upaniṣads* look for what is beyond the past and the future.[33] They seek infinitive *(Bhūman)*[34] and plinitude *(Pūrṇa)* which they find symbolized more in space than in Time : the atmosphere, infinite space *(Ākāśa)* is present also in the innermost chambers of the heart *(Hṛdyākāśa)*.

As regards the Indian conception of Time i.e., *Kālavāda*, the *Maitrāyaṇa Upaniṣad* compares Time with *Brahman.*[35] Time and Not Time are called the two bodies of the Brahman. So the reflections of the *Maitrāyaṇa Upaniṣad* on the Transitory nature of everything existing led to the consideration that Time is substance identical with *Brahman*.

Bṛhadāraṇyaka Upaniṣad makes Time only proceed from Brahman through *Ākāśa,* "Space", which is often equal to the other.[36] *Śvētāśvatara Upaniṣad* mentions *Kāla* with Nature, *Svabhāva,* Necessity, *Niyati,* Chance, *Yādṛcca,* the Fundamental substance, *Bhūtāni, Yōni* and *Puruṣa* among the possible primary principles which may have caused the world to come into existence.[37]

However, *Upaniṣads* distinguish two aspects of Brahman, of Universal Being : "the corporal and the incorporeal, the mortal and the immortal, the fixed *(Sthita)* and the mobile etc."[38] *Maitrāyaṇa Upaniṣad* establishing this by-polarity of Universal Being on the plane of Time, distinguishes two forms *(Dvirūpe)* of Brahman i.e., the aspects of the 'two natures' of a single essence *(Tadēkaṃ),* as 'Time and Without-Time' *(Kālaścakāśca).*[39] In other words, Time, Eternity are the two aspects of the same principle : in Brahman, *nune fluens* and the *nune stans* coincide. The *Maitrāyaṇa Upaniṣad* continues : What precedes the Sun is Without Time *(Akāla)* and Undivided *(Akāla);* but what begins with the Sun is Time which has parts *(Sakāla)* and its form is the year.....". The expression "what precedes the Sun" may be interpreted cosmologically as relating to the epoch which preceded the creation for in the intervals between the *Mahāyugas* or *Kalpas*, during the Great Cosmic Nights, Time no longer exists—but its application is above all metaphysical and esoteriological : it refers to the paradoxical situation of him who obtains illumination, who becomes a *Jīvanmukta,* who is "delivered in this life" and thereby transcends Time in the sense that he no longer participates in it.

Chāndōgya Upaniṣad declares that for the sage, the illumined one, the Sun remains motionless. "But after having risen to the Zenith it (the Sun) will never rise or set again. It will remain alone in the centre *(Ēkala Ēva Madhya Sthita).* Whence this verse : There (in the transcendent world of Brahman) it has never set and never set and never risen..... ?

It neither rises nor sets; for him who knows the doctrine of Brahman it is in the heavens once for all *(Sakṛt)*". Here, of course, we have a concrete image of transcendence at the Zenith, i.e., at the summit of the celestial vault, at the "Centre of the world".....where a cutting.....across the planes, a communication between the three cosmic zones, is possible ..... the Sun (Time remains immobile for "him who knows"; the *nune fluens* paradoxically transferred into a *nune stans*. Illumination, understanding, accomplish the miracle of an escape from Time.

In *Upaniṣads* the instant of illumination is lightened to the lightening flash-space "in lightening flash the truth.[40] *Brahman* is understood suddenly like a lightening flash."[41] *Kēna Upaniṣad* on *Brahman,* the Universal spirit, the absolute being is conceived as Transcending and on the source and the foundation of everything that is manifested in Time :[42] "Lord of what he has been and will be, is both today and tomorrow."

In *Maitrāyaṇa Upaniṣad* meditation on Time is recommended as a method for getting beyond Time to the timeless reality. The *Maitrāyaṇa Upaniṣad* mentions Time as one of the principal forms of the supreme, immortal, unembodied *Brahman*.[43] It is as a form of *Brahman* that time is to be meditated on. The Supreme Brahman is not in the sphere of what is seen, and is without shape. So the mind can't dissolve in it directly. One of the indirect ways of mind-dissolution is the contemplation of Time as the image of *Brahman*. Meditation implies imaginative substitution. It involves necessarily a make-believe. But it is useful in that it leads us through the image to the real. The image may be external or internal, gross or subtle. As one progresses in meditation, one makes the image more and more subtle of the most images of *Brahman,* Time is pre-eminent. The fruit of Time-meditation is not to cling to Time as if it were ultimately real, but to transcend it. As the *Maitrāyaṇa Upaniṣad* declares, "He who worships Time as *Brahman,* from him Time withdraws afar".[44]

In the *Muṇḍaka Upaniṣad* it is stated in (Mantra 1-7) that the whole world is created from the absolute, whatever existed in the past, whatever exists now, whatever will exist in future is nothing but AUM and what is beyond these three Times also AUM.

Birth and death are mysterious events and have intimate relation with the concept of Time.[45] In *Kaṭha Upaniṣad* there is a story in which *Nacikēta* has put the question to *Yama* (the God of death) as to where a dead man goes after his death. *Yama* has given him a long reply, but the substance of the whole discourse is contained in the following two stanzas :

यदेवेह तदामृत् यदामृत् तदान्विहि।
मृत्योः स मृत्युमाप्नोति य इहि नानेव पश्चतिः॥ (२.१.१०)

Yadēvēha Tadāmṛt Yadāmṛt Tadānvihi ।
Mṛtyōḥ Sa Mṛtyumāpnōti Ya Ihi Nānēva Paścatiḥ ॥
(2-1-10)

That is what is here is there and what is there is here. Whosoever sees differences or duality experiences the cycles of birth and death.

मनसैवेद माप्तव्यत्नेद नानास्ति किंचन।
मृत्योः स मृत्युं गच्चति य इह नानेव पश्यति॥ (२.१.११)

Manasaivēda Māptavyatnēda Nānāsti Kiñcana ।
Mṛtyōḥ Sa Mṛtyuṃ Gaccati Ya Iha Nānēva Paśyati ॥
(2-1-11)

That is this (all pervading spirit of *Brahma*) can be realized only by a pure mind. Here no differences exist.

Whoever sees differences or duality experiences the cycles of birth and death.

The *Maitrāyaṇa Upaniṣad* mentions Time as one of the principle forms of the supreme, immortal, unembodied *Brahman*. It is a form of *Brahman* the Time is to be meditated on. The *Upaniṣad* also mentions Time as a cosmic power and an absolute principle of fate. "From Time flow beings, through Time they grow old; in Time they are destroyed. Time is that is amorphous assumes a shape."[46]

Śvētāśvatara Upaniṣad attacks the *Kālavāda* as being materialistic and atheistic. "Some sages say that the cause of the world is nature, others say Time. They are mistaken, it is the power of good which causes the Brahman's wheel to turn in this world."[47] The *Upaniṣad* further says, 'Time is not an independent reality, it is the Lord who is the 'knower and creator of Time,' and Time is his instrument."[48] This *Upaniṣad* emphasizes the transcendence of the Lord vis-a-vis Time : 'He is the origin—He is beyond three fold time—He is higher, He is other than the tree (of the world), other than time and forms......'[49] It is also mentioned in the *Upaniṣad* that He, *Rudra,* reprocesses the world at end of Time.[50] Thus the *Upaniṣads* more or less deal with the transcendence of Time rather than with the nature of Time.

Brahmasūtras

Brahmasūtras are based on the philosophy of the *Upaniṣads*. As the question of time is left indefinite in the *Upaniṣads,* in the *Brahmasūtras* also it is in the same position. In many *Sūtras,* the fifth case *Vibhakti* is used, thus indicating that something has come from something else, yet the origin of Time has not been clearly stated. Several reasons have been given to show that everything which exists now, which existed before and which exists in future is nothing but the

Absolute. With regard to creation, it is stated in the fifth *Sūtra, Ikṣata Naśabdaṃ* i.e., inert nature is not the cause of Universe because it has no power of observation. Therefore, the Absolute Brahman is the only cause, because it has the power of observation. This statement is made on the basis of what is said in the *Upaniṣads*.

In the *Brahmasūtras* real change or *Pariṇāmavāda* has been negated on the basis of what is stated in the *Upaniṣads*, but the question of Time does not appear to have been keenly taken up for inquiry.

The emphasis in the *Brahmasūtras* is laid on correcting the knowledge of division *Dvaita Jñāna* and he cause of this knowledge is given as ignorance. Space and time are both shown to be products of ignorance *Avidya* but no further explanation is forthcoming. It is stated in the last part of the *Brahmasūtras* that God can create only world or all the worlds, but a liberated soul can't create anything. This shows that the question of time has not been properly considered therein.

Mahābhārata

Mahābhārata, one of the great epics is dated by the moderns to the Post-Vēdic period. If consists of 18 books and about one lakh of couplets, and believed that it was written by *Vēdavyāsa*.

In *Mahābhārata* different views have been held about time. The predominating one certainly seems to be that of an insurmountable fate. The most frequent quotation in *Mahābhārata* which is attributed to the *Kālavādin* is the following :[51]

"Time ripens beings, time enfolds creatures.
Time keeps watch when all are asleep.

Time is hard to overcome."

"But no one here on earth knows Him in whom Time is Ripened."

Time is the Lord who works change in beings—that which can't be understood and that from which there is no return. Time is the destiny (flux *gati*) of everything; if one does not follow it, where can we go? Whether you try to flee from it or remain motionless, you can't escape from time. The five senses can't grasp it. Some say that *(Kāla)* is fire, and others that it is the Lord of the creatures *(Prajāpati)* some conceive of time as a season, others as a month, a day or even an instant—There are those who say that it is the hour *(Muhūrta)* : but that which is uniquely one has many forms. Time must be acknowledged as that which controls everything that exists.[52]

In the *Mahābhārata,* time is perceived as indivisible and omnipotent, beyond divisible and measurable time.

Time is the cause of all, it is Time that creates and destroys,[53] that it binds men by its links,[54] and causes the joys and sufferings of men, regardless of their actions.[55] According to this conception man is purely and simply delivered upto fate, and his actions and efforts are powerless to alter his lot. In the end, the destructive aspect dominates. It is Time which hastens the progress of all beings towards dissolution.[56] Time is compared to an ocean where one can see neither the other shore nor any island of refuge.[57] Time becomes the great destructive power, sometimes synonymous with death.[58]

The frequently expressed idea that time 'matures' or ripens beings means simply that it leads to old age and ultimately to death. 'Even the Gods are subject to this impermance.'[59]

In the *Mahabhārata* the pre-eminent place of Time is expressed as—"Beyond the spirit is intelligence, beyond intelligence Great Time, (but) beyond Time is Lord *Viṣṇu,* from whom proceeds all the Universe."[60]

There are certain parts in the text of *Mahābhārata* deal with theories which aim at explaining the world on the basis of *Svabhāva* (self-existence or Nature), *Niyati* (Fate or Destiny), *Yādṛcca* (Chance or some invisible force of cosmic will), and *Kāla* (Time). Their promulgation is associated with the demons—the enemies of the Gods. The prominent features of these theories are found in the passages of the *Mahābhārata,* especially of the *Mōkṣa-Dharma* section which contains the largest number of philosophical passages in the Mahābhārata.

In the Chapter XII, 172, in reply to *Yudhiṣṭhira's* query as to how a man can be free from grief and can attain the highest state *(Paramagatiḥ) Bhīṣma* narrates an ancient dialogue between *Prahlāda,* the king of demons and *Ajagara Muni.* In that dialogue, *Prahlāda,* struck by *Ajagara's* wonderful equanimity, child-like innocence and freshness of mind (3-4) asked the sage the secret of his mental equipoise. In reply to that question, the sage *Ajagara* discourses on his doctrine of *Svabhāva.* He says :

"All activities, O Prahlāda, are due to and find consummation in Nature *(Svabhāva).*"[61]

There is no cause why things arise, grow, decay and perish.[62] Strengthened by this knowledge, I neither feel elated nor dejected (10). All things, big or small, have an end (13). The luminaries in the sky are also seen falling down when their Time comes (*Yathākālaṃ* 17). Whatever fare, rich or coarse, I get by chance (*Yādṛcca* 19) is welcome to me. I pass days together without anything to eat. Whatever has come my way by chance (24), I never reject, nor do I go after what is difficult to get. I see that pleasure and pain, fortune and misfortune, love and hate, life and death are all ordained by fate (*Vidhiniyata* 30). Fortified by this knowledge, I live the *Ajagara* life of vow in order to restrain my thirsty and unsteady mind (33).[63]

The characterisitc feature of the doctrine of *Svabhāva* as described in the above passage is that all events, occurrences and effects in this world and human life came about without any cause. So also from the words in the above passage, it will be obvious that the doctrine of *Svabhāva* (Nature) is closely associated with or closely impunges on the doctrines of Time, Fate and Chance.

The Chapter XII, 219 sets forth the doctrine of Destiny or Fate. *Bhīṣma* narrates the conversation between *Indra* and *Namuci. Namuci,* the demon king, was defeated. Still he remained unruffled and calm like an ocean (2). Asked by Indra as to how he took his present disaster, *Namuci* propounds his doctrine of Destiny : "There is one ruler without any second. He rules over man from the time he is in the womb. Under his instructions, and as directed by him I carry on like water running down the slope."[64]

In this passage in the beginning, destiny appears to be cast in the theistic guise of a personal Ruler *(Śāstā)*; but the whole tenor of the description following after it suggests that what is meant is some mythopoeic power of *Fate* or Destiny endowed with a will, before which, human effort and will have no place. It should also be noticed that, in the above passage, Destiny stands associated with *'Paryāya'* i.e., Time i.e., the Time-process.

The two Chapters XII, 217 and 220 contain a very powerful advocacy of the doctrine of Time at the hands or rather from the mouth of *Bali*—the king of the demons. In reply to *Yudhiṣṭhira's* question as to how those who are hit hard by Time behave, *Bhīṣma* narrates in this connection an ancient dialogue between *Indra* and *Bali : Bali* once reigned supreme. It was he who once made the sunrise and set, the rains fall and the fire burn (216.5). He was, however, vanquished by the Gods and passed his days in obscurity, reduced to the life of an ass grazing on husks and chaff. *Indra,* mounted on his elephant *Airāvata*, sought him out and asked him how he took his present. Then Bali proceeds to expand his doctrine of Time. He says : I do not grieve, Oh *Indra.* For I know that everything has to perish in course of Time. The events in Time are fleeting and evanescent (217.5). Time, with its power seizes everything (19). The ultimate fate of things is destruction (9). A man who kills and the man who is killed—both have already been struck down before by Time (14) One burnts what has already been burnt, kills what has already been killed (20). 'Earth, wind, space, water and light the fifth—those are the source of creatures (17). Time presides over these. Time seizes everthing; it is Time that dispenses things; it is Time that holds everything. Human effort is vain (217-25). Prosperity or otherwise is all brought about by Time (36). My turn to rise and be aggressive is now over; it is now my time to suffer (39 & 44). It is Time that puts everything in its position; it is Time that ripes or cooks everthing (39). All things are subject to Time whether you run or stop in one place, Time will not leave you (51). Time can't be grasped by the senses. It consists of seasons, months, fortnights, days, morning, noon, evening and moments (52-53). Oh *Indra,* you are now so powerful; but you will also be 'put out' by Time (55). Time takes me away in my time and will take that I had conquered you; it is on account of Time that you have now conquered me (220.35). Time first overtook me and is now running after you (220.68). Thousands of kings of the demons have passed away in the course of Time. It is difficult to transgress Time (220.41). See, there is this dark terrible Man of Time, hard to overcome, who has bound me like an animal with a rope (220.82).

Towards the end, *Indra* compliments *Bali* to the effect that *Bali* is the philosopher of Time and that he has the closest and the most intimate knowledge of Time (220-105).

Thus, according to the doctrine of *Kāla,* everything in this world has been pre-ordained victory and defeat, prosperity and poverty, the coming together of the elements and their disintegration. Man is but a helpless creature in the hands of this dreadful Time.

Besides the above passages, there are also some other passages in the *Śāntiparvaṃ* of the *Mahābhārata* which associate the doctrines of *Svabhāva* and *Kāla* with materialism. It is said that thinkers who philosophize over the primary material elements posit the ultimate principle of *Svabhāva* or of *Kāla*.[65] According to these thinkers, all things, whether sentient or insentient, are the product of the Elements which come together or dissolve by nature i.e., by *Svabhāva* or under the influence of Time or *Kāla*.

It will be seen from the above references in the *Mahābhārata* that the doctrines of *Svabhāva, Niyati, Yādṛcca,* and *Kāla* preach a strong determinism, leaving no scope for human effort. The conceptions underlying these different doctrines appear to overlap, and are almost identical with one another. Whether it be the inherent nature of things *(Svabhāva* or *Naturalism),* or the inevitable Destiny *(Niyati* or *Fatalism),* or some mysterious will or whim immanent in things *(Yādṛcca* or *accidentalism)* or whether it be Time *(Kāla)* which presides over the present, past and future of things-all these agree in that they pre-determine the course of events in this world and in man's life—in fact the whole garnet of existence from birth, growth and decay to final dissolution. The two ideas fundamental to all these doctrines are the impermanence of the world and man, and the negation of human will and effort. The thorough going determinism of these doctrines is based on materialism, according to which everything in the world including human life is the product of the Material Elements (Earth, Water, Fire, Air and Space) which come together and go off at the behest of *Svabhāva, Kāla* etc. The eloquent way in which these powers have been described in the above passages suggests that each power is conceived as a mythopoeic personification invested with a will of its own, governing in its supreme sway, the whole course of the world and of human life. It is also, further significant that these doctrines of *Svabhāva* and *Kāla* presented in these passages were advocated by or associated with the demons—the enemies of the Gods.

Further *Kāla* in many passages of *Mahābhārata,* represents the Time or the moment of the fructification of man's *Karma.* Thus, when any calamity, disaster or death befalls, or a man acts at a particular moment in a particular way, it is said that such occurrences or events occur under the impulse of *Kāla,* implying thereby that particular occurrence or event represents the fructifying moment of the past *Karman. 'Kālacōdita'* (impelled by Time) is one of the stock phrases of the *Mahābhārata* in such contexts.

Bhagavadgīta

The *Bhagavadgīta* has been regarded as the most important scripture of the Post-Vēdic period. The aim of the *Bhagavadgīta* was to show a definite and clear path to *Arjuna* in the great war of *Mahābhārata* by *Lord Kṛṣṇa* and *Arjuna* after hearing *Gīta* admitted that all his doubts have been removed.

In *Bhagavadgīta,* Lord *Kṛṣṇa* manifesting himself to *Arjuna* as a Cosmic God declares : "I am the Time which, in progressing, destroys the world".[66] So in *Bhagavadgīta, Kāla* is conceived as an aspect of the God-head, especially its terrible aspect, and again as calculation of Terrestrial.[67] In the great war of *Mahābhārata, Arjuna* requested *Śrī Kṛṣṇa* to place his chariot in such a position that he might be able to see the armies of both the sides. When this was done, he saw that so many persons would be killed in the battle for the sake of a piece of land and began to tremble and said to *Śrī Kṛṣṇa* that he did not like to fight. The persons who had gathered there for the battle were relatives and friends of *Arjuna* and *Śrī Kṛṣṇa* also, but *Śrī Kṛṣṇa* saw the coming events with different knowledge. In his vision Time was unimportant. He wanted *Arjuna* to change the nature of his observation and said that the soul is neither born nor does it die. It does not come into existence at the time of birth and doesn't cease to exist when killed in a battle. It is eternal and unborn.

Here an attempt is made by the Lord *Kṛṣṇa* to change the nature of observation. When the observation is changed Time can't remain as before. When the personality shifts, Time doesn't mean the same thing. The question of Time depends on the relationship which holds between a human-observer and his finite level.

Lord *Kṛṣṇa* not only gave spiritual instructions to *Arjuna* (as stated in the 11th Chapter of the *Gīta*) but also gave a new power of observation by which *Arjuna* saw that the whole Universe with its innumerable forms existed as one whole being. In this vision, *Arjuna* said, he saw neither the end, nor middle, nor beginning of anything. Besides he said there is no fundamental space. Therefore, Lord *Kṛṣṇa* advises *Arjuna* not to look to the result of his action, but the correct method of acting by right thinking and right observation. The functions of Time and Space have infinite number of dimensions.

In the fourth chapter of the *Gīta,* the Lord tells *Arjuna* that he gave a new method of observation to *Vivasvān* (The Sun God)—"I am the beginning, the middle and the end of all living beings".[68]

In the eleventh Chapter the Lord says 'I am the Time', (*Kālōsmin* 11-30). This means when an observer creates new Time, he will observe a new creation. This means also that the previous creation as seen by the previous Time must go. Thus the Time comes from the observer and if *Arjuna* changes his method of observation he will observe as if all the warriors are already dead. This means giving up all previous mental, vital and physical habits. When a new habit has been formed it will create a new vision of the world. Then there will be new laws according to the nature of field. This field is neither true nor false (13-12). Therefore, the field must be transcended by attending to the knower of the field.

If the Universe was created so that Time runs from the epoch of creation then it must be concluded also that the Universe was created as a point singularity. This is in accordance with what the Lord *Kṛṣṇa* says that when the observer sees that all has come from and has remained in the one and the all is the spreading of the observations of the one, he attains the state of right observation (13-30). As Time is not fundamental this field has no real beginning nor real end (15-2). *Arjuna* replied—"But you are born now and *Vivasvān* was born in the beginning of the creation. How then was it possible for you to give a new vision to *Vivasvān*". Lord *Kṛṣṇa* said, "I had many births and you also had many births, I know them all but you can't know owing to the limitation of your power of observation; my birth and my actions are all divine that is based on new method of observation".

The Lord says, "the cycle of birth and death brings innumerable pains. Only he who attains to my stage will be delivered from them (8-15). Even if a man obtains the *Brahma* stage he will have to come back because *Brahma* is also subject to Time (8-16). His method of observation created new Time for him. Yet, he remains subject to birth and death. But, to those who are absolutely devoted to me I give new power of knowledge whereby they come to me".

Therefore, in *Bhagavadgīta,* it is *Kṛṣṇa* who is identified with Time in its indestructible but destroying aspect that is the eternal devours Time.

Śrīmad Bhāgavata

Bhāgavata is considered to be one of the best philosophical works in India. In its introduction, there is a dialogue between four *Sanatkumārs* (who remain together and who do not grow old) and *Nārada*. The cause of their not growing old is given in stanzas 2-48. That is, *Nārada* says to *Sanatkumārs* that as they are continuously reciting the words, "we surrender to God", old age due to Time does not have any effect on them. This means that Time can be changed by transferring and attention from the temporary embodied self to the eternal disembodied Supreme spirit. But, what is Time itself has not been clearly stated.

At some places in the *Bhāgavata* it is stated, that Time has come out of God [(2-5-21) and (3-6-21)]. At other places, it is stated that Time does not exist in God [(2-2-17) and (2-9-107)]. This means that the original composer of *Bhāgavata* is not quite sure about the nature of Time.

A statement about Time appears in Canto 3 chapter 11 of *Bhāgavata.* It begins with one *Tṛṭi* which is equal to the Time taken by the Sun's light in passing three molecules *(Aṇu),* each molecule consisting of three atoms *(Paramāṇu)*. On this basis, further combination of Time has been arranged. This is in connection with the outward Time. But it also stated that the Time which is spread in a fundamental particle *(Aṇu)* is equal to that particle. Further combinations of Time such as *Ghaḍi, Prahar, Day, Pakṣa, Month* and *Yuga* are based on this original Time. In the last Canto No. 12 it is stated that *Yugas* or cycles of Time are not true. When evil nature begins in a man then *Kali Yuga* or particular cycle of Time begins in him.

Therefore, Time is not fundamental. It is formless yet makes an appearance when some motion takes place in the nature. This is clearly stated in the stanza 3-10-11. That is, which appears as a change in the original nature as cosmic intelligence but which has no form or substance in Time. By that means God creates himself as the world.

In the fifth Canto, there is a story of the creation of seven worlds by the movement of the Sun's chariot. Each world is called a cycle of Time. Time is formed by the movement of the Sun but no further explanation is given.

A wonderful story about Time is given in Canto 9, chapter 3 as follows :

A king *Rēvat* had a daughter by name *Rēvati.* When she came of age, the king went to the world of *Brahma* (creator) to ask him to whom his daughter should be married. At that time in the court of *Brahma,* singing and dancing were going on. Therefore, the king had to wait there for a short while. When the dancing and singing were over, the king asked *Brahma* to which king his daughter should be married. *Brahma* said—"during the short time that you are here, several *Yugas* (ages) have passed in your world and all the kings that you had seen are dead. Now a new cycle of Time i.e., *Yuga* has commenced and *Balarāma* who is a form of God has taken birth, you should therefore give your daughter to *Balarāma* in marriage". The king followed the advice of *Brahma.*

This means that every world has its own Time, yet Time is considered as an objective reality for the persons living in the same world. No explanation is given here, why Time should be different in different worlds.

In Canto 10, while introducing the king *Kamsa, (Śukadēv*

says to *Parīkṣit)* that he was *Kālanēmi* (cycle of Time) in his previous birth (10-2-68). He had great enemity towards God *Viṣṇu* and His followers, therefore he sent several of his evil-minded servants to kill *Kṛṣṇa* and his followers, but they all failed and *Kamsa* was finally killed by *Kṛṣṇa*. This means, in other words that *Kṛṣṇa* had victory over Time.

In the beginning of the Tenth Canto (10-17), it is stated that God is giving life to all beings by remaining inside the body and is killing all beings by assuming the form of Time outside the body. This means that those who look to only outside things will feel the effect of Time.

Purāṇas

Purāṇas echo the conception of Time as divinity and often attempt to integrate it in their respective theologies. The *Brahmavaivarta Purāṇa* mentions Great Time as an instrument of knowledge and hence of deliverance from the bonds of *Māya*. The discussion of the myth of the Time in this *Purāṇa* elucidates its function in destroying God Indra's vanity of pride of being in an exalted position.

After the victory over the dragon *Vṛtra, Indra* decides to build and embellish the residence of the Gods. *Viśvakarman,* the divine architect, labours for an year and succeeds in constructing a magnificient palace. But Indra is not satisfied : he wishes to make it still larger and more splendid, without its equal in the world. Exhausted with his effort, *Viśvakarman* complains to *Brahma,* the Creator God. *Brahma* promises to help him and intervenes with *Viṣṇu,* the Supreme Being, of whom *Brahma* himself is only a simple instrument. *Viṣṇu* undertakes to *Indra* back to his senses.

One fine day, *Indra* in his palace receives the visit of a ragged boy. It is *Viṣṇu* himself, who has assumed this aspect to humiliate the king of the Gods without immediately revealing his identity. He calls *Indra* "my child" and speaks to him of the innumerable *Indras* who have inhabited innumerable Universes up to his time.....

The life and kingship of an *Indra* endure 71 *eons* (a cycle, a *Mahāyuga,* consists of 12,000 divine years or 4,32,000 years) and when 28 *Indras* have expired, one day and night of *Brahma* have elapsed. But the existence of one *Brahma,* measured in such *Brahma* days and nights, is only 108 years. *Brahma* follows *Brahma;* one sinks the next arises; the endless series can't be told. There is no end to the number of those *Brahmas*—to say nothing of *Indras.*

But the Universes side by side at any given moment, each harbouring a *Brahma* and an *Indra;* who will estimate the manner of these? Beyond the farthest vision, crowding outer space, the Universes come and go, like an innumerable hosts. Like delicate boats they float on the fathomless, pure waters that form the body of *Viṣṇu.* Out of every hairpore of that body a Universe bubbles and breaks. Will you presume to count them? Will you number the Gods in all those worlds—the worlds present and worlds past?

As the boy speaks, a procession of ants, has made its appearance in the great hall of the palace. Drawn up in a column four yards wide, they parade across the floor. The boy perceives them pauses, and then, seized with amazement, breaks out in a sudden laugh. "Why do you laugh?" *Indra* asks him.

And the boy replies : "I saw the ants, O *Indra,* filing in long parade. Each was once an *Indra.* Like you, each by virtue of pious deeds once ascended to the rank of a king of Gods. But now, through many rebirths each has become again an ant. This army is an army of former Indras.....".

This revelation brings home to *Indra* the vanity of his pride and ambitions. He recalls the admirable architect *Viśvakarman,* rewards him royally, and abandons forever his project of enlarging the palace of the Gods.

So *Indra* realizes his transitory and temporal character and the falsity of his pride and position. This story reveals to *Indra* about Great Time, Mythical Time which is the true source of all cosmic beings and events.

Indra feels enlightened, transcends his empirical status and surrenders to the mythical sacred Time, which is equivalent to a revelation of the ultimate reality.

In *Viṣṇu Purāṇa* it is stated "*Kāla* is without beginning and without end, ageless, Omni-present and Supremely free, that it is the Great Lord"[69] which echoed the conception of Time as a divinity i.e., a Cosmic Power and an absolute principle. The *Bhāgavata Purāṇa* also emphasizes the same as "Time being infinite, cause the end; being without a beginning, creates the beginning, the immutable".[70] Further on the subject of the divine immanence, the *Bhāgavata Purāṇa* says : The Lord penetrates all existences by his own power *(Ātma Māya)* within he takes the form of the spirit *(Puruṣa)*, on the outside he takes the form of Time *(Kāla-Rūpa)*."[71] The cosmological role of *Kāla* is described in *Brahma Purāṇa* as—"The Time exists in a latent state during the dissolution of the world, and is awakened by the God at the moment of the recreation".[72]

Regarding *Kālavāda, Viṣṇu Purāṇa* mentions Time as one of the four shapes of the highest essence, the other three being *Pradhāna, Puruṣa* and *Vyakta.*[73] On the other hand, *Bhāgavata Purāṇa* makes *Viṣṇu* act through *Kāla* and occasionally *Viṣṇu* is called *Kāla* or *Kāla* appears as an independent element besides *Viṣṇu.*[74]

Śiva Purāṇa recognizes three levels of Time. In its first stage, Time is not different from *Śiva,* it is eternal. In its second stage, it becomes the power of *Śiva, Śiva* being the innermost essence *(Ātman)* of Time. *Śiva* rules the Universe by means of Time. In the third stage, Time is considered as an united principle being the product of *Māya,* Cosmic

Illusion. Only in this last stage is Time divided up and causes succession, duration and limitation. So, the transcendent absolute aspect of Time has been transferred into the sphere of God, and its empirical aspect into the sphere of *Māya* that which veils Reality.

In both the *Śaivite* and *Vaiṣṇavite* theologies any reality which is not identical with God-though very frequently identified with him—becomes his power of *Śakti*. Because of its cosmological importance Time is one of the God's earliest powers : *Kālaśakti,* his instrument in the creation, preservation and destruction of the Universe.

Yet *Kāla* is, and remains, more closely linked to *Śiva* than to *Viṣṇu* and it is its destructive aspect which dominates all other cosmological aspects. *Śiva* himself is called *Mahākāla,* Great Time, meaning death. The destructive Goddess *Kāḷi* is the female counterpart of the God *Rudra-Śiva,* who is identified with Time.

Ṣaṭ Darśanas

The *Ṣaṭ Darśanas* form an important part of Hindu thought. The *Darśanas* are given in the form of *Sūtras* or aphorisms which are sought and definite. The ideas of *Avidya, Māya, Puruṣa* and *Jīva* are common to all these schools of thought. They all believe in the creation, maintenance and dissolution of the world. Excepting the *Pūrva Mīmāmsa* the other systems aim at *Mōkṣa* or the release of the soul from rebirths.

Nyāya-Vaiśēṣhika

Time is conceived in the *Nyāya-Vaiśēṣika* system as a unique, all-pervading and eternal substance. It is the static background against which events happen and from which they derive their chronological order. It possesses no specific physical quality like colour and thus can't be an object of external perception. Neither is it perceivable internally, for the mind has no jurisdiction over external or non-physical objects independently of a physical sense-organ. The question naturally arises : What is the source of our knowledge that time exists? The *Vaiśēṣika* answers that the knowledge of time is arrived at by a series of inferences. The notions of priority *(Paratva)* and posteriority *(Aparatva)*, of simultaneity *(Yaugapādya)* and succession *(Ayaugapādya)*, and of quickness *(Kṣipratva)* and slowness *(Ciratva)* constitute the grounds *(Liṅga)* of inference of existence of Time.[75]

It is one in number, Omni-present in dimension, individual in character, and possesses the qualities of conjunction and disjunction. The past of an individual is the time characterised by its destruction, the future that characterised by its precedent non-existence, the present is the time whose future existence is destroyed and whose own destruction is about to come.

In the *Nyāyasūtra*,[76] the question of the existence of the present time is discussed in answer to the objection that, when an object falls, we know only the time through which it has fallen and the time through which it still will fall. The reply is that without a present, there can be no perception and no knowledge, and past and future would have no meaning or existence. In the *Vaiśēṣika Sūtra* (II-6-9; V, 2, 26; VII, 1, 25; 2-22), there is further the pregnant doctrine that Time is a cause for transient things in which it exists, but not for eternal things in which it is not found. This is a clear recognition of the fact that the eternal substances do not exist in time, while their qualities and motions have only existence there with all the products. But the term cause must not be understood in the sense that the *Vaiśēṣika* adopted the popular view that time was a great Cosmic Power which caused movement of things;[77] this is wholly contradictory to the view of the *Sūtra*, which never attributes the origin of motion to time, as well as to the harmonious tradition of the school.

Time is a cause only in the loose sense of that term, which is affected in the school, namely, as one of the many conditions which are necessary to the existance of a thing; it belongs to the category of general instrumental cause.[78] As opposed either to the specific instrumental cause *(Kāraṇa)* or the inherent or non-inherent causes.

The nature of the conjunction which results in the conceptions of priority and posteriority is made clear by the stock example of the school. If we say, 'Here now is a jar', we operate with a conjunction of a special character, that of the Sun and the jar; this conjunction is not like that of material objects or their parts, and it is due to some reality which must be inferred, and is inferred as Time. So again if we make a youth our starting point, the cognition of priority is produced in the case of an old man whose birth precedes that of the youth by many revolutions of the Sun.[79] These motions of the Sun are the conditions which mark the divisions of Time, such as moments, months and days. It is these conditions which render Time apparently manifold instead of one as it really is, and help to create the impression held by some of the *Nyāya* school that time was, as claimed by the Buddhists, merely a series of moments, a view, which, as we have seen, the *Nyāya* and *Vaiśēṣika* repudiated as a general principle, but adopted with modifications in their theory of the character of all non-external existence, in the shape of the theory that every quality and action lasts for three moments only. There remains one obvious difficulty in regarding time itself as eternal, when its presence in eternal substances is denied. But in both cases the term 'eternal' denotes that which has no cause saves itself and really exists, and which in the temporal terms which we cannot avoid in use can only be described as eternal, unless we frame the concept—though it can have no concrete meaning for us-of that which is not in time at all.

Time—therefore, is regarded as a fundamental reality which is the basis of our Time knowledge with which all transient existence is bound up.

Sāṅkhya-Yōga System

The *Sāṅkhya-Kārika* of *Īśvarakṛṣṇa* is the earliest available work of the *Sāṅkhya* system. It does not say anything on the concept of *Kāla*. Among his commentaries only the *Sāṅkhya-Tatva-Kaumudi* incidentally touches it in the following passage :

"According to the *Vaiśēṣikas* time is one (Indivisible) and hence it can't admit of such divisions as past, present and future. They attribute these divisions to certain "accidents" (adventitious conditions) : but according to the *Sāṅkhyācāryas* these same "accidents" themselves may be regarded as the basis of the notions of the future, present and past and there is no need for the postulating of an intervening entity as 'Time'. This is the reason why Time is not accepted as a distinct entity".[80]

From the *Upaniṣadic* period onwards both Time-succession and duration and the Universe of the act *(Karma)* are divested of their value, and at the end of this process it is the doctrine of the cycle of existences *(Saṃsāra)* which gives rise to a negative conception of Time. The meta-physical schools which have liberation (Mōkṣa) as their goal tend in theory to deny Time any real value, and seek to achieve a state of existence which is beyond Time—to use a *Yōga* term "the cessation of mental states" *(Yōga Citta Vṛtti-Nirōdha,* V.S.I.1), one of which is Time.

In order to affirm the relativity of Time, these schools then envisage Time as an intellectual concept having no counterpart in any 'real thing'. This almost psychological reduction of Time is expressed, for instance, in the *Yōga Vaśiṣtha,* which attempts to demonstrate the unreality of instants and cosmic ages. (Y.Vās III, 20, 29). It says that, depending on the mental state of the subject, an instant may appear as a *Kalpa (aeon)* or contrariwise, an *aeon* may be experienced as a single instant (Y.Vās III.60, 21; III, 103, 14 etc.). In short, Time perse does not exist. He who is absorbed in meditation knows neither day nor night (Y.Vās. III.60, 25-6) and in the end, self-knowledge, illumination encompasses all the future in an instant (Y.Vās. III.60, 171).

In his *Yōgasūtra, Patañjali* recommends 'meditation on the instant and the succession of instants in order to attain knowledge born of discernment' (Y.S.III, 52).

The *Vyāsa-Bhāṣya* on the *Yōgasūtras* of *Patañjali,* the commentaries of *Vācaspati Miśra (Tattva-Vaiśāradi)* and *Vijñānabhikṣu (Yōga-Vārtika)* on the *Bhāṣya* deal with the concept of *Kāla* at some length. Thus the *Vyāsa-Bhāṣya* (III.51) says : 'As the atom is a substance in which minuteness reaches its limit, so a moment is a decision of time in which minuteness reaches its limit. Or a moment is that much of time which an atom takes in leaving the position in space it occupies and reaching the next point. The succession of moment is the non-cessation of the flow thereof. The moments and their collection do not fall into a collection of actual things. The *Muhūrta,* the day and night are all aggregates of mental conceptions. This Time which is not a substantive reality in itself, but is only a mental concept, and which comes into the mind as a piece of verbal knowledge only, appears to people whose minds are given to out-going activities *(Vyuttatadarśanānāṃ)* as if it were an objective reality.

The moment falls under the head of reality and is maintained by succession. This succession consists in the sequence of moments. The yogis who know time, call this by the name of Time.[81]

Further, two moments can't co-exist. There can be no

succession of two co-existent moments. It is impossible. The uninterrupted sequence of the first moment and of the one, which follows, is what is called succession. For this there is but one moment existing in the present; the antecedent and postecedent moments do not exist; therefore there can't be any collection of them.

Further, the moments that have passed and those that have yet to come, should be described as existing in consequence of Universal change in evolution. For this reason the whole world undergoes change every moment and all these characteristics are relatively established in that one moment of time.

Vijñānabhikṣu in his *Yōga-Vārtika* points out that though Time is defined as *Buddhi-Nirmāṇa,* yet it may be taken as real. For it only means that Time has no real (Objective) existence apart from the 'moment'. The latter is a real being identical with the unit of change of the *Guṇas* in phenomena *(Yōga Vārtika* on III, 51). But as Dr. Seal points out, even this is real only for our empirical (relative) consciousness *(Vyuttatadarśana),* which intuits the relation antecedence and sequence into the evolving Reals *Guṇas* in the stage of 'Empirical intuition' *(Savicāra Nirvikalpa Prajña),* on the other hand, apprehends the Reals as they are, without the imported empirical relations of Space, Time and Causality.[82]

From the above it is clear that the *Sāṅkhya Yōga* accepts *Kāla*, not as a distinct eternal entity, like the *Vaiśēṣikas,* but only as the one present moment which is identical with the unit of change of the *Guṇas.*

REFERENCES

1. आयुर्विश्वायुः परि पासति त्वा पूषा त्वा पातु प्रपथे पुरस्तात्।
यत्रासते सुकृतो यत्र ते ययुस्तत्र त्वा देवः सविता दधातु॥

ऋग्वेद, १०.१७.४.

Āyurviśvāyuḥ Pari Pāsati Tvā Pūṣā Tvā Pātu Prapathē Purastāt I
Yatrāsatē Sukṛtō Yatra Tē Yayustatra Tvā Dēvaḥ Savitā Dadhātu II

Ṛgvēda, X.17.4.

2. वयश्चितू ते पतत्रिणो द्विपंचतुष्पदर्जुनि।
उषः प्रारन्नृतूंरनु दिवो अन्तेभ्यस्परि॥

ऋग्वेद, १.४.३.

Vayaścitū Tē Patatriṇō Dvipañcatuṣpadarjuni I
Uṣaḥ Prārannṛtūṃranu Divō Antēbhyaspari II

Ṛgvēda, I.4.3.

को अग्निमीट्टे हविषा घृतेन सुचा यजाता ऋतुभिर्धुवेभिः।
कस्मै देवा आ वहानाशु होम को मंसते वीतिहोत्रः सुदेवः॥

ऋग्वेद, १.८४.१८.

Kō Agnimīṭṭē Haviṣā Ghṛtēna Sucā Yajātā Ṛtubhirdhuvēbhiḥ I
Kasmai Dēvā Ā Vahānāśu Hōma Kō Maṃsatē Vītihōtraḥ Sudēvaḥ II

Ṛgvēda, 1.84.18.

3. Raimundo Panikkar, "Time and History in the Tradition of India : Kāla and Karma," in Culture and Time, ed.L. Gaslet et al, Paris, UNESCO Press, 1976, pp. 64.

4. यस्य ते विश्वा भुवनानि केतुना प्र चेरते नि च विशन्ते अक्तुभिः।
अनागास्त्वेन हरिकेश सूर्याह्नाह्ना नो वस्यसावस्यसोदिहि॥

ऋग्वेद, १०.३७.९.

Yasya Tē Viśvā Bhuvanāni Kētunā Pra Cēratē Ni Ca Viśantē Aktubhiḥ I
Anāgāstvēna Harikēśa Sūryāhnāhnā Nō Vasyasāvasyasōdihi II

Ṛgveda, X.37.9.

शं नो भव चक्षसा शं नो अह्ना शं भानुना शं हिमा शं घृणेन।
यथा शमध्वञ्छमसद् दुरोणे तत् सूर्य द्रविणं धेहिचित्रम्॥१०॥

Śaṃ Nō Bhava Cakṣasā Śaṃ Nō Ahnā Śaṃ Bhānunā Śaṃ Himā Śaṃ Ghṛṇēna I
Yathā Śamadhvañchamasad Durōṇē Tat Sūrya Draviṇaṃ Dhēhi Citram II10II

अस्माकं देवा उभयाय जन्मने शर्म यच्छत द्विपदे चतुष्पदे।
अदत् पिबदूर्जयमानमाशितं तदस्मे शं योररपो दधातन॥११॥

Asmākaṃ Dēvā Ubhayāya Janmanē Śarma Yacchata Dvipadē Catuṣ padē I

Adat Pibadūrjayamānamāśitaṃ Tadasmē Śaṃ Yōrarapō Dadhātana//11//

यद्वो देवाश्चकृम जिह्वया गुरु मनसो वा प्रयुती देवहेळनम्।
अरावा यो नो अभि दुच्छुनायते तस्मिन् तदेनो वसवो नि धेतन॥१२॥

Yadvō Dēvāścakṛma Jihvayā Gurū Manasō Vā Prayutī Dēvahēḷanam/
Arāvā Yō Nō Abhi Ducchunāyatē Tasmin Tadēnō Vasavō Ni Dhētana//12//

5. अग्निना रयिमश्नवत् पोषमेव दिवेदिवे। यशसं वीरवत्तमम्॥

ऋग्वेद, १.१.३.

Agninā Rayimaśnavat Pōṣamēva Divēdivē/ Yaśasaṃ Vīravatta-mam//

Ṛgvēda 1.1.3.

विश्वाहा त्वा सुमनसः सुचक्षसः प्रजावन्तो अनमीवा अनागसः।
उद्यन्तं त्वा मित्रमहो दिवेदिवे ज्योग्जीवाः प्रति पश्येम सूर्य॥

ऋग्वेद, १०.३७.७

Viśvāhā Tvā Sumanasaḥ Sucakṣasaḥ Prajāvantō Anamīvā Anāgasaḥ/
Udyantaṃ Tvā Mitramahō Divēdivē Jyōgjīvāḥ Prati Paśyēma Sūrya// Ṛgvēda X.37.7.

6. इयं वेदिः परो अन्तः पृथिव्या अयं यज्ञो भुवनस्य नाभिः।
अयं सोमो वृष्णो अश्वस्य रेतो ब्रह्मायं वाचः परमं व्योम॥

ऋग्वेद, १.१६४.३५.

Iyaṃ Vēdiḥ Parō Antaḥ Pṛthivyā Ayaṃ Yajñō Bhuvanasya Nābhiḥ/
Ayaṃ Sōmō Vṛṣṇō Aśvasya Rētō Brahmāyaṃ Vācaḥ Paramaṃ Vyōma// Ṛgvēda, I.164.35.

7. सहस्रशीर्षा पुरुषः सहस्राक्षः सहस्रपात्।
स भूमिं विश्वतो वृत्वा त्यतिष्ठद्दशाङ्गुलम्॥

ऋग्वेद, १०.९०.१.

Sahasraśīrṣā Puruṣaḥ Sahasrākṣaḥ Sahasrapāt/
Sa Bhūmiṃ Viśvatō Vṛtvā Tyatiṣṭhaddaśāṅgulam// Ṛgvēda, X.90.1.

पुरुष एवेदं सर्वं यद्भूतं यच्च भव्यम्।
उतामृतत्वस्येशानो यदन्नेनातिरोहति॥२॥

Puruṣa Ēvēdaṃ Sarvaṃ Yadbhūtaṃ Yacca Bhavyam/
Utāmṛtatvasyēśānō Yadannēnātirōhati//2//

एतावनस्य महिमा तो ज्यायाँश्च पूरुषः।
पादोस्य विश्वा भूतानि त्रिपादस्यामृतंदिवि॥३॥
Ētāvanasya Mahimā Tō Jyāyāmśca Pūruṣaḥ।
Pādōsya Viśvā Bhūtāni Tripādasyāmṛtandivi॥3॥
त्रिपादूर्ध्व उदैत् पुरुषः पादोस्येहा भवत् पुनः।
ततो विष्वङ् व्यक्रामत् साशनानशने अभि॥४॥
Tripādūrdhva Udait Puruṣaḥ Pādōsyēhā Bhavat Punaḥ।
Tatō Viṣvaṅ Vyakrāmat Sāśanānaśanē Abhi॥4॥
तस्मादिराळजायत विराजो अधि पूरुषः।
स जातो अत्यरिच्यत पश्चाद्भूमिमथो पुरः॥५॥
Tasmādvirāḷajāyata Virājō Adhi Pūruṣaḥ।
Sa Jātō Atyaricyata Paścādbhūmimathō Puraḥ॥5॥
यत् पुरुषेण हविषा देवा यज्ञमतन्वत।
वसंतो अस्यासीदाज्यं ग्रीष्म इध्मः शरद्धविः॥६॥
Yat Puruṣēṇa Haviṣā Dēvā Yajñamatanvata।
Vasantō Asyāsīdājyaṃ Grīṣma Idhmaḥ Śaraddhaviḥ॥6॥
तं यज्ञं बर्हिषि प्रौक्षन् पुरुषं जातमग्रतः।
तेन देवा अयजंत साध्या ऋषयश्च ये॥7॥
Taṃ Yajñaṃ Barhiṣi Praukṣan Puruṣaṃ Jātamagrataḥ।
Tēna Dēvā Ayajanta Sādhyā Ṛṣayaśca Yē॥7॥
तस्माद्यज्ञात! सर्वहुतः संभृतं पृषदाज्यम्।
पशून् ताँश्चक्रे वायव्या नारण्यान् ग्राम्याश्च ये॥८॥
Tasmādyajñāt Sarvahutaḥ Sambhṛtaṃ Pṛṣadājyam।
Paśūn Tāmścakrē Vāyavyā Nāraṇyān Grāmyāśca Yē॥8॥
तस्माद्यज्ञात् सर्वहुत ऋचः सामानि जज्ञिरे।
छन्दांशि जज्ञिरे तस्माद्यजुस्तस्मादजायत॥९॥
Tasmādyajñāt Sarvahuta Ṛcaḥ Sāmāni Jajñirē।
Chandāṃśi Jajñirē Tasmādyajustasmādajāyata॥9॥
तस्मादश्वा अजायन्त ये के चोभयादतः।
गावो ह जज्ञिरे तस्मात् तस्माज्जाता अजावयः॥१०॥
Tasmādaśvā Ajāyanta Yē Kē Cōbhayādataḥ।
Gāvō Ha Jajñirē Tasmāt Tasmājjātā Ajāvayaḥ॥10॥
यत् पुरुषं व्यदधुः कतिधा व्यकल्पयन्।
मुखं किमस्य कौ बाहू का ऊरु पादा पच्येते॥११॥

Yat Puruṣaṃ Vyadadhuḥ Katidhā Vyakalpayan I
Mukhaṃ Kimasya Kau Bāhū Kā Ūrū Pādā Ucyētē II11 II
ब्राह्मणोस्य मुखमासीद्बाहू राजन्यः कृतः।
ऊरू तदस्य यद्वैश्यः पद्भ्यां शूद्रो अजायत॥१२॥
Brāhmaṇōsya Mukhamāsīdbāhū Rājanyaḥ Kṛtaḥ I
Ūrū Tadasya Yadvaiśyaḥ Padbhyāṃ Śūdrō Ajāyata II12 II
चंद्रमा मनसो जातश्चक्षोः सूर्यो अजायत।
मुखादिन्द्रश्चाग्निश्च प्राणाद्वायुरजायत॥१३॥
Candramā Manasō Jātaścakṣōḥ Sūryō Ajāyata I
Mukhādindraścāgniśa Prāṇādvāyurajāyata II13 II
नाभ्या आसीदन्तरिक्षं शीर्ष्णो द्यौः समवर्तत।
पद्भ्यां भूमिर्दिशः श्रोत्रात् तथा लोकाँ अकल्पयन्॥१४॥
Nābhyā Āsīdantarikṣaṃ Śīrṣṇō Dyauḥ Samavartata I
Padbhyāṃ Bhūmirdiśaḥ Śrōtrāt Tathā Lōkāṁ Akalpayan II14 II
सप्तास्यासन् परिधयस्त्रिः सप्त समिधः कृताः।
देवा यद्यज्ञं तन्वाना अभध्नन् पुरुषं पशुम्॥१५॥
Saptāsyāsan Paridhayastriḥ Sapta Samidhaḥ Kṛtāḥ I
Dēvā Yadyajñaṃ Tanvānā Abhadhnan Puruṣaṃ Paśum II15 II
यज्ञेन यज्ञमयजंत देवास्तानि धर्माणि प्रथमान्यासन्।
ते ह नाकं महिमानः सचन्त यत्र पूर्वे साध्याः सन्ति देवाः॥१६॥
Yajñēna Yajñamayajanta Dēvāstāni Dharmāṇi Prathamānyāsan I Tē Ha Nākaṃ Mahimānaḥ Sacanta Yatra Pūrvē Sādhyāḥ Santi Dēvāḥ II16 II

8. *Ibid.*
9. वहिष्ठेभिर्विहरन्यासि तन्तु मवव्ययन्नसितं देव वस्म।
दविध्वतो रश्मयः सूर्यस्य चर्मेवावाधुस्तमो अप्स्वान्तः॥

ऋग्वेद, ४.१३.४.

Vahiṣṭhēbhirviharanyāsi Tantu Mavavyayannasitaṃ Dēva Vasma I Davidhvatō Raśmayaḥ Sūryasya Carmēvāvāvādhustamō Apsvāntaḥ II Ṛgvēda.IV.13.4.

सुप्रतीके वयोवृधा यह्वी ऋतस्य मातरा। दोषामुषासमीमहे॥

ऋग्वेद, ५.५.६.

Supratīkē Vayōvṛdha Yahvī Ṛtasya Mātarā I Dōṣāmuṣāsamīmahē II
Ṛgvēda.V.5.6.

यो विद्यात् सूत्रं विततं यस्मिन्नोताः प्रजा इमाः।
सूत्रं सूत्रस्य यो विद्यात् स विद्याद् ब्राह्मणं महत्॥

अथर्ववेद, १०.८.३७.

Yō Vidyāt Sūtraṃ Vitataṃ Yasminnōtāḥ Prajā Imāḥ//
Sūtraṃ Sūtrasya Yō Vidyāt Sa Vidyād Brāhmaṇaṃ Mahat//

Atharvavēda, X.8.37.

10. तन्त्रमेके युवती विरूपे अभ्याक्रामं वयतः षण्मयूखम्।
प्रान्या तन्तूस्तिरते धत्ते अन्या नाप वृञ्जाते न गमातो अन्तम्॥

अथर्ववेद, १०.७.४२.

Tantramēkē Yuvatī Virūpē Abhyākrāmaṃ Vayataḥ Ṣaṇmayūkham/
Prānyā Tantūstiratē Dhattē Anyā Nāpa Vṛññātē Na Gamātō Antam//

Atharvavēda, X.7.42.

तयोरहं परिनृत्यन्त्योरिव न वि.जानामि यतरा परस्तात्।
पुमानेनद् वयत्युद्गृणत्ति पुमानेनद् वि जभारधि नाके॥४३॥
Tayōrahaṃ Parinṛtyantyōriva Na Vi Jānāmi Yatarā Parastāt/
Pumānēnad Vayatyudgṛṇatti Pumānēnad Vi Jabhāradhi Nākē//43//

11. साध्वपांसि सनता न उक्षिते उषासानक्ता वय्येव रण्विते।
तन्तुं ततं संवयन्ति समीची यज्ञस्य पेशः सुदुघे पयस्वती॥

ऋग्वेद, २.३.६.

Sādhvapāṃsi Sanatā Na Ukṣitē Uṣāsānaktā Vayyēva Raṇvitē/
Tantuṃ Tataṃ Saṃvayanti Samīcī Yajñasya Pēśaḥ Sudughē Payasvatī//

Ṛgvēda, II.3.6.

यो यज्ञो विश्वतस्तन्तुभिस्तत एकशतं देवकर्मेभिरायतः।
इमे वयन्ति पितरो य आययुः प्र वयाप वयेत्यासते तते॥

ऋग्वेद, १०.१३०.१.

Yō Yajñō Viśvatastantubhistata Ēkaśataṃ Dēvakarmēbhirāyataḥ/
Imē Vayanti Pitarō Ya Āyayuḥ Pra Vayāpa Vayētyāsatē Tatē//

Ṛgvēda, X.130.1.

पुमाँ एनं तनुत उत् कृणत्ति पुमान् वि तत्ने अधि नाके अस्मिन्।
इमे मयूखा उप सेदुरू सदः सामानि चक्रुस्तसराण्योतवे॥२॥
Pumāṁ Ēnaṃ Tanuta Ut Kṛṇatti Pumān Vi Tatnē Adhi Nākē Asmin/
Imē Mayūkhā Upa Sēdurū Sadaḥ Sāmāni Cakrustasarāṇyōtavē//2//

12. आ ते पितर्मरुतां सुम्नमेतु मा नः सूर्यस्य संदृशो युयोथाः।
अभि नो वीरो अर्वति क्षमेत प्र जायेमहि रुद्र प्रजाभिः॥

ऋग्वेद, २.३३.१.

Ā Tē Pitarmarutām Sumnamētu Mā Naḥ Sūryasya Sandṛśō Yuyōthāḥ l
Abhi Nō Vīrō Arvati Kṣamēta Pra Jāyēmahi Rudra Prajābhiḥ ll

Ṛgvēda.II.33.1.

त्वादत्तेभी रुद्र शंतमेभिः शतं हिमा अशीय भेषजेभिः।
व्यस्मद् द्वेषो वितरं व्यंहो व्यमीवाश्चातयस्वा विषूचीः॥२॥

Tvādattēbhī Rudra Śantamēbhiḥ Śataṃ Himā Aśīya Bhēṣajēbhiḥ l
Vyasmad Dvēṣō Vitaraṃ Vyaṃhō Vyamīvāścātayasvā Viṣūcīḥ ll2ll

श्रेष्ठो जातस्य रुद्र श्रियासि तवस्तमस्तवसां वज्रबाहो।
पर्षि णः पारमंहसः स्वस्ति विश्वा अभीती रपसो युयोधि॥३॥

Śrēṣṭhō Jātasya Rudra Śriyāsi Tavastamastavasāṃ Vajrabāhō l
Parṣi Ṇaḥ Pāramaṃhasaḥ Svasti Viśvā Abhītī Rapasō Yuyōdhi ll3ll

मा त्वा रुद्र चुक्रुधामा नमोभिर्मा दुष्टुती वृषभ मा सहूती।
उन्नो वीराँ अर्पय भेषजेभिर्भिषक्तमं त्वा भिषजां शृणोमि॥४॥

Mā Tvā Rudra Cukrudhāmā Namōbhirmā Duṣṭutī Vṛṣabha Mā Sahūtī l
Unnō Vīrāṁ Arpaya Bhēṣajēbhirbhiṣaktamaṃ Tvā Bhiṣajāṃ Śṛṇōmi ll4ll

हवीमभिर्हवते यो हविर्भिरव स्तोमेभी रुद्रं दिषीय।
ऋदूदरः सुहवो मा नो अस्यै बभ्रुः सुशिप्रो रीरधन्मनायै॥५॥

Havīmabhirhavatē Yō Havirbhirava Stōmēbhī Rudraṃ Diṣīya l
Ṛdūdaraḥ Suhavō Mā Nō Asyai Babhruḥ Suśiprō Rīradhanmanāyai ll5ll

उन्मा ममन्द वृषभो मरुत्वान् त्वक्षीयसा वयसा नाधमानम्।
घृणीव च्छायामरपा अशीयाविवासेयं रुद्रस्य सुम्नम्॥६॥

Unmā Mamanda Vṛṣabhō Marutvān Tvakṣīyasā Vayasā Nādhamānam l
Ghṛṇīva Cchāyāmarapā Aśīyāvivāsēyaṃ Rudrasya Sumnam ll6ll

क्व स्य ते रुद्र मृळयाकुर्हस्तो यो अस्ति भेषजो जलाषः।
अपभर्ता रपसो दैव्यस्याभी नु मा वृषभ चक्षमीथाः॥७॥

Kva Sya Tē Rudra Mṛḷayākurhastō Yō Asti Bhēṣajō Jalāṣaḥ l
Apabhartā Rapasō Daivyasyābhī Nu Mā Vṛṣabha Cakṣamīthāḥ ll7ll

प्र बभ्रवे वृषभाय श्वितीचे महो महीं सुष्टुतिमीरयामि।
नमस्या कल्मलीकिनं नमोभिर्गृणीमसि त्वेषं रुद्रस्य नाम॥८॥

Pra Babhravē Vṛṣabhāya Śvitīcē Mahō Mahīṃ Suṣṭutimīrayāmi I Namasyā Kalmalīkinaṃ Namōbhirgṛṇīmasi Tvēṣaṃ Rudrasya Nāma II8II

स्थिरेभिरङ्गैः पुरुरूप उग्रो बभ्रुः शुक्रेभिः पिपिशे हिरण्यैः।
ईशानादस्य भुवनस्य भूरेर्न वा उ योषद् रुद्रादसुर्यम्॥९॥

Sthirēbhiraṅgaiḥ Pururūpa Ugrō Babhruḥ Śukrēbhiḥ Pipiśē Hiraṇyaiḥ I

Īśānādasya Bhuvanasya Bhūrērna Vā U Yōṣad Rudrādasuryam II9II

अर्हन् बिभर्षि सायकानि धन्वार्हन् निष्कं यजतं विश्वरूपम्।
अर्हन्निदं दयसे विश्वमभ्वं न वा ओजीयो रुद्र त्वदस्ति॥१०॥

Arhan Bibharṣi Sāyakāni Dhanvārhan Niṣkaṃ Yajataṃ Viśvarūpam I Arhannidaṃ Dayasē Viśvamabhvaṃ Na Vā Ōjīyō Rudra Tvadasti II10II

स्तुहि श्रुतं गर्तसदं युवानं मृगं न भीममुपहत्नुमुग्रम्।
मृळा जरित्रे रुद्र स्तवानोन्यं ते अस्मन्नि वपन्तु सेनाः॥११॥

Stuhi Śrutaṃ Gartasadaṃ Yuvānaṃ Mṛgaṃ Na Bhīmamupahatnu-mugram I

Mṛḷā Jaritrē Rudra Stavānōnyaṃ Tē Asmanni Vapantu Sēnāḥ II11 II

कुमाराश्चेत् पितरं वन्दमानं प्रति नानाम रुद्रोपयन्तम्।
भूरेर्दातारं सत्पतिं गृणीषे स्तुतस्त्वं भेषजा रास्यस्मे॥१२॥

Kumārāśvēt Pitaraṃ Vandamānaṃ Prati Nānāma Rudrōpayantam I

Bhūrērdātāraṃ Satpatiṃ Gṛṇīṣē Stutastvaṃ Bhēṣajā Rāsyasmē II12II

या वो भेषजा मरुतः शुचीनि या शंतमा वृषणो या मयोभु।
यानि मनुरवृणीता पिता नस्ता शं च योश्च रुद्रस्य वश्मि॥१३॥

Yā Vō Bhēṣajā Marutaḥ Śucīni Yā Śantamā Vṛṣaṇō Yā Mayōbhu I

Yāni Manuravṛṇītā Pitā Nastā Śaṃ Ca Yōśca Rudrasya Vaśmi II13II

परि णो हेती रुद्रस्य वृज्याः परि त्वेषस्य दुर्मतिर्मही गात्।
अव स्थिरा मघवद्भ्यस्तनुष्व मीढ्वस्तोकाय तनयाय मृळ॥१४॥

Pari Ṇō Hētī Rudrasya Vṛjyāḥ Pari Tvēṣasya Durmatirmahī Gāt I Ava Sthirā Maḍhavadbhyastanuṣva Mīḍhvastōkāya Tanayāya Mṛḷa II14II

एवा बभ्रो वृषभ चेकितान यथा देव न हृणीषे न हंसि।
हवनश्रुन्नो रुद्रेह बोधि बृहद् वदेम विदथे सुवीराः॥१५॥

Ēvā Babhrō Vṛṣabha Cēkitāna Yathā Dēva Na Hṛṇīṣē Na Haṃsi/
Havanaśrunnō Rudrēha Bōdhi Bṛhad Vadēma Vidathē Suvīrāḥ//15//

13. कालो अश्वो वहति सप्तरश्मिः सहस्राक्षो अजरो भूरिरेताः।
तमा रोहन्ति कवयो विपश्चितस्तस्य चक्रा भुवनानि विश्वा॥

अथर्ववेद, १९.५३.१.

Kālō Aśvō Vahati Saptaraśmiḥ Sahasrākṣō Ajarō Bhūrirētāḥ/
Tamā Rōhanti Kavayō vipaścitastasya Cakrā Bhuvanāni Viśvā//

Atharvavēda, XIX.53.1.

सप्त चक्रान् वहति काल एष सप्तास्य नाभीरमृतं न्वक्षः।
स इमा विश्वा भुवनान्यञ्जत् कालः स ईयते प्रथमो नु देवः॥२॥
Sapta Cakrān Vahati Kāla Ēṣa Saptāsya Nābhīramṛtaṃ Nvakṣaḥ/
Sa Imā Viśvā Bhuvanānyañjat Kālaḥ Sa Īyatē Prathamō Nu Dēvaḥ//2//

पूर्णः कुम्भोधि काल आहितस्तं वै पश्यामो बहुधा नु सन्तः।
स इमा विश्वा भुवनानि प्रत्यङ्कालं तमाहुः परमे व्योमन्॥३॥
Pūrṇaḥ Kumbhōdhi Kāla Āhitastaṃ Vai Paśyāmō Bahudhā Nu Santaḥ/
Sa Imā Viśvā Bhuvanāni Pratyaṅkaālaṃ Tamāhuḥ Paramē Vyōman//3//

स एव सं भुवनान्याभरत् स एव सं भुवनानि पर्यैत्।
पिता सन्नभवत् पुत्र एषां तस्माद् वै नान्यत् परमस्ति तेजः॥४॥
Sa Ēva Saṃ Bhuvanānyābharat Sa Ēva Saṃ Bhuvanāni Paryait/
Pitā Sannabhavat Putra Ēṣāṃ Tasmād Vai Nānyat Paramasti Tējaḥ//4//

कालोमूं दिवमजनयत् काल इमाः पृथिवीरुत।
काले ह भूतं भव्यं चेषितं ह वि तिष्ठते॥५॥
Kālōmūṃ Divamajanayat Kāla Imāḥ Pṛthivīruta/
Kālē Ha Bhūtaṃ Bhavyaṃ Cēṣitaṃ Ha Vi Tiṣṭhatē//5//

कालो भूतिमसृजत काले तपति सूर्यः।
काले ह विश्वा भूतानि काले चक्षुर्वि पश्यति॥६॥
Kālō Bhūtimasṛjata Kālē Tapati Sūryaḥ/
Kālē Ha Viśvā Bhūtāni Kālē Cakṣurvi Paśyati//6//

काले मनः काले प्राणः काले नाम समाहितम्।
कालेन सर्वा नन्दन्त्यागतेन प्रजा इमाः॥७॥

Kālē Manaḥ Kālē Prāṇaḥ Kālē Nāma Samāhitam I
Kālēna Sarvā Nandantyāgatēna Prajā Imāḥ II7II
काले तपः काले ज्येष्ठं काले ब्रह्म समाहितम्।
कालो ह सर्वस्येश्वरो यः पितासीत् प्रजापतेः॥८॥
Kālē Tapaḥ Kālē Jyēṣṭhaṃ Kālē Brahma Samāhitam I
Kālō Ha Sarvasyēśvarō Yaḥ Pitāsīt Prajāpatēḥ II8II
तेनेषितं तेन जातं तदु तस्मिन् प्रतिष्ठितम्।
कालो ह ब्रह्म भूत्वा बिभर्ति परमेष्ठिनम्॥९॥
Tēnēṣitaṃ Tēna Jātaṃ Tadu Tasmin Pratiṣṭhitam I
Kālō Ha Brahma Bhūtvā Bibharti Paramēṣṭhinam II9II
कालः प्रजा असृजत कालो अग्रे प्रजापतिम्।
स्वयंभूः कश्यपः कालात् तपः कालादजायत॥१०॥
Kālaḥ Prajā Asṛjata Kālō Agrē Prajāpatim I
Svayambhūḥ Kaśyapaḥ Kālāt Tapaḥ Kālādajāyata II10II
कालादापः समभवन् कालाद् ब्रह्म तपो दिशः।
कालेनोदेति सूर्यः काले नि विशते पुनः॥
अथर्ववेद, १९.५४.१.
Kālādāpaḥ Samabhavan Kālād Brahma Tapō Diśaḥ I
Kālēnōdēti Sūryaḥ Kālē Ni Viśatē Punaḥ II
Atharvavēda, XIX.54.1.
कालेन वातः पवते कालेन पृथिवी मही।
द्यौर्मही काल आहिता॥२॥
Kālēna Vātaḥ Pavatē Kālēna Pṛthivī Mahī I
Dyaurmahī Kāla Āhitā II2II
कालो ह भूतं भव्यं च पुत्रो अजनयत् पुरा।
कालादृचः समभवन् यजुः कालादजायत॥३॥
Kālō Ha Bhūtaṃ Bhavyaṃ Ca Putrō Ajanayat Purā I
Kālādṛcaḥ Samabhavan Yajuḥ Kālādajāyata II3II
कालो यज्ञं समैरयद् देवेभ्यो भागमक्षितम्।
काले गन्धर्वाप्सरसः काले लोकाः प्रतिष्ठिताः॥४॥
Kālō Yajñaṃ Samairayad Dēvēbhyō Bhāgamakṣitam I
Kālē Gandharvāpsarasaḥ Kālē Lōkāḥ Pratiṣṭhitāḥ II4II
कालेयमङ्गिरा देवोथर्वा चाधि तिष्ठतः।
इमं च लोकं परमं च लोकं पुण्यांश्च लोकान् विधृतीश्च पुण्याः।
सर्वाल्लोकानभिजित्य ब्रह्मणा कालः स ईयते परमो नु देवः॥५॥

Kālēyamaṅgirā Dēvōtharvā Cādhi Tiṣṭhataḥ I
Imaṃ Ca Lōkaṃ Paramaṃ Ca Lōkaṃ Puṇyāṃśca Lōkān Vidhṛtīśca Puṇyāḥ I
Sarvāllōkānabhijitya Brahmaṇā kālaḥ Sa Īyatē Paramō Nu Dēvaḥ II5 II

14. Translation by M. Bloomfield—'Hymns of the Atharva Vēda', S.B.E. (1987), p. 224-225.
15. Atharvavēda, XIX, 53, 1-8, by trans. by L. Renou, in : Hymnes speculatif du Vēda (verses 9 and 10 have been deleted), Paris, 1956. The quotation is a UNESCO Secretariat translation, from French.
16. Op. Cit., pp. 224-225.
17. कालेयमङ्गिरा देवोथर्वा चाधि तिष्ठतः।
इमं च लोकं परमं च लोकं पुण्यांश्च लोकान् विधृतीश्च पुण्याः।
सर्वांल्लोकानभिजित्य ब्रह्मणा कालः स ईयते परमो नु देवः॥
अथर्ववेद, १९.५४.५.

Kālēyamaṅgirā Dēvōtharvā Cādhi Tiṣṭhataḥ I
Imaṃ Ca Lōkaṃ Paramaṃ Ca Lōkaṃ puṇyāṃśca Lōkān Vidhṛtīśca Puṇyāḥ I
Sarvāllōkānabhijitya Brahmaṇā Kālaḥ Sa Īyatē Paramō Nu Dēvaḥ II
Atharvavēda, XIX.54.5.

18. कालो ह भूतं भव्यं च पुत्रो अजनयत् पुरा।
कालादृचः समभवन् यजुः कालादजायत॥
अथर्ववेद, १९.५४.३.

Kālō Ha Bhūtaṃ Bhavyaṃ Ca Putrō Ajanayat Purā I
Kālādṛcaḥ Samabhavan Yajuḥ Kālādajāyata II
Atharvavēda, XIX.54.3.

19. कालेन वातः पवते कालेन पृथिवी मही।
द्यौर्मही काल आहिता॥
अथर्ववेद, १९.५४.२.

Kālēna Vātaḥ Pavatē Kālēna Pṛthivī Mahī I
Dyaurmahī Kāla Āhitā II
Atharvavēda, XIX.54.2.

20. काले मनः काले प्राणः काले नाम समाहितम्।
कालेन सर्वा नन्दन्त्यागतेन प्रजा इमाः॥
अथर्ववेद, १९.५३.७.

Kālē Manaḥ Kālē Prāṇaḥ Kālē Nāma Samāhitam।
Kālēna Sarvā Nandantyāgatēna Prajā Imāḥ॥

Atharvavēda, XIX.53.7.

21. कालो भूतिमसृजत काले तपति सूर्यः।
काले ह विश्वा भूतानि काले चक्षुर्वि पश्यति॥

अथर्ववेद, १९.५३.६.

Kālō Bhūtimasṛjata Kālē Tapati Sūryaḥ।
Kālē Ha Viśvā Bhūtāni Kālē Cakṣurvi Paśyati॥

Atharvavēda, XIX.53.6.

22. O.G. Von Wesendonck, 'The Kālavāda and the Zervanite System,' Journal of the Royal Asiatic Society of Great Britain and Ireland, 1931, p. 56.
23. *Ibid.*, p. 56.
24. *Ibid.*, p. 56.
25. *Ibid.*, p. 56.
26. रोहितः कालो अभवद् रोहितोग्रे प्रजापतिः।
रोहितो यज्ञानां मुखं रोहितः स्वराभरत्॥

अथर्ववेद, १३.२.३९.

Rōhitaḥ Kālō Abhavad Rōhitōgrē Prajāpatiḥ।
Rōhitō Yajñānāṃ Mukhaṃ Rōhitaḥ Svarābharat॥

Atharvavēda, XIII.2.39.

27. जानाम्यहँ शेवधिरित्यनित्यं न ह्यध्रुवैः प्राप्यते हि ध्रुवं तत्।
ततो मया नाचिकेतश्चितोग्निरनित्यैर्द्रव्यैः प्राप्तवानस्मि नित्यम्।

कठोपनिषद्, २.१०.

Jānāmyahaṁ Śēvadhirityanityaṃ Na Hyadhruvaiḥ Prāpyatē Hi Dhruvaṃ Tat।
Tatō Mayā Nācikētaścitōgniranityairdravyaiḥ Prāptavānasmi Nityam॥ *Kaṭhōpaniṣad, II.10.*

कामस्याप्तिं जगतः प्रतिष्ठां क्रतोरनन्त्यमभयस्य पारम्।
स्तोममहदुरुगायं प्रतिष्ठां दृष्ट्वा धृत्या धीरो नचिकेतोत्यस्राक्षीः॥११॥
Kāmasyāptiṃ Jagataḥ Pratiṣṭhāṃ Kratōranantyamabhayasya Pāram।
Stōmamahadurugāyaṃ Pratiṣṭhāṃ Dṛṣṭvā Dhṛtyā Dhīrō Nacikētōtya Srākṣīḥ॥11॥

28. अस्तमित आदित्ये याज्ञवल्क्य चन्द्रमस्यस्तमिते शान्तेग्नौ किं ज्योतिरेवायं पुरुष इति वागेवास्य ज्योतिर्भवतीति वाचैवायं ज्योतिषास्ते पल्ययते कर्म

कुरुते विपल्येतीति तस्माद्वै सम्राडपि यत्रस्व पाणिर्न विनिर्ज्ञायतेथ यत्र वागुच्चरत्युपैव तत्रन्येती त्येव मेवै तद्याज्ञवल्क्य।

बृहदारण्यक उपनिषद्, ६.३.५.

Astamita Ādityē Yājñavalkya Candramasyastamitē Śāntēgnau Kiṃ Jyōtirēvāyaṃ Puruṣa Iti Vāgēvāsya Jyōtirbhavatīti Vācaivāyaṃ Jyōtiṣāstē Palyayatē Karma Kurutē Vipalyētīti Tasmādvai Samrāḍ-api Yatrasva Pāṇirna Vinirjñāyatētha Yatra Vāguccaratyupaiva Tatranyētī Tyēva Mēvai Tadyājñavalkya।

Bṛhadāraṇyaka Upaniṣad, VI.3.5.

29. अथातो व्व्रतमीमाँसा। प्प्रजापतिर्ह कर्म्माणि ससृजे तानि सृस्टान्यन्योन्येनास्पर्द्धन्त व्वदिष्याम्येवाहमिति व्वाग्दद्धे द्रक्ष्याम्यहमिति चक्षुः श्रेष्याम्यहमिति श्रोत्रमेवमन्यानि कर्म्माणि यथाकर्म्म॥

बृहदारण्यक उपनिषद्, १.३.३०.

Athātō Vvratamīmāṁsā। Pprajāpatirha Karmmāṇi Sasṛjē Tāni Sṛṣṭānyanyōnyēnā Sparddhanta Vvadiṣyāmyēvāhamiti Vvāgdaddhē Drakṣyāmyahamiti Cakṣuḥ Śrēṣyāmyahamiti Śrōtramēvamanyāni Karmmāṇi Yathākarmma॥

Bṛhadāraṇyaka Upaniṣad, I.3.30.

तानि मृत्युः श्रमो भूत्वोपयेमे। तान्यान्पोत्तान्याप्त्वा मृत्युरववारुन्द्ध तस्माच्छ्राम्यत्येव व्वाक्छ्राम्यति चक्षुः श्राम्यति श्रोत्रमथेममेव नाप्नोद्योयम्मध्यमः प्प्राणः॥३१॥

Tāni Mṛtyuḥ Śramō Bhūtvōpayēmē। Tānyānpōttānyāptvā Mṛtyura-vavārunddha Tasmācchrāmyatyēva Vvākchrāmyati Cakṣuḥ Śrāmyati Śrōtramathēmamēva Nāpnōdyōyammadhyamaḥ Pprāṇaḥ॥31॥

तानि ज्ञातुन्दद्ध्ररे। अयम्वै नः श्रेष्ठो यः सञ्चरंश्चासंञ्चरंश्च न व्व्यथतेथो न रिष्यति हन्तास्यैव सर्व्वे रूपम्भवामेति त एतस्यैव सर्व्वे रूपमभवँस्तस्मादेतएतेनाख्यायन्ते प्प्राणा इति तेन ह वाव तत्कुलमाख्यायते यस्मिन्कुले भवति य एवम्वेद य उ हैवम्विदा स्पर्द्धते अनुशुष्य हैवान्ततो म्म्रियते इत्यध्यात्मम्॥३२॥

Tāni Jñātundaddhrarē। Ayamvai Naḥ Śrēṣṭhō Yaḥ Sañcaraṃ-ścāsaññcaraṃśca Na Vvyathatēthō Na Riṣyati Hantāsyaiva Sarvvē Rūpambhavāmēti Ta Ētasyaiva Sarvvē Rūpamabhavamstasmā-dētaētēnākhyāyantē Pprāṇā Iti Tēna Ha Vāva Tatkulamākhyāyatē Yasminkulē Bhavati Ya Ēvamvēda Ya U Haivamvidā Sparddhatē Anuśuṣya Haivāntatō Mmriyatē Ityadhyātmam॥32॥

30. अथाधिदेवतम्। ज्वलिष्याम्येवाहमित्यग्निर्दद्धे तपस्याम्यहमित्यादित्यो भास्याम्यहमिति चन्द्रमा एवमन्या देवता यथादेवतँ स यथैषाम्प्राणानाम्मध्यमः प्प्राण

एवमेतासान्देवता नाम्वायुम्लोंचन्ति ह्यन्या देवता न व्वायुः सैषानस्तमिता देवता यद्वायुः॥

बृहदारण्यक उपनिषद्, १.३.३३.

Athādhidēvatam। Jvaliṣyāmyēvāhamityagnirddaddhrē Tapasyāmyahamitya Ādityō Bhāsyāmyahamiti Candramā Ēvamanyā Dēvatā Yathādēvatam Sa Yathaiṣām Prāṇānām Madhyamaḥ Pprāṇa Evamētāsāndēvatānāmvā Yumlōñcanti Hyanyā Dēvatā Na Vvāyuḥ Saiṣānastamitā Dēvatā Yadvāyuḥ॥

Bṛhadāraṇyaka Upaniṣad, I.3.33.

31. अथैष श्लोको भवति। यतश्चोदेति सूर्योस्तं यत्र च गच्छतीति प्राणादद्वा एष उदेति प्राणेस्तमेति तन्देवाश्चक्रिरे धर्म्मं स एवाद्य स उ श्च इति यद्वा एतेमुह्यद्ध्रियन्त तदेवाप्यद्य कुर्वन्ति तस्मादेकमेव व्व्रतञ्चरेत्प्राण्याञ्चैवापान्याच्च नेन्मा पाप्मा म्मृत्युराप्नवदिति यद्यु चरेत्समापिपयिषेत्तेनो एतस्यै देवतायै सायुज्यँ सलोकताञ्जयति य एवम्वेद॥

बृहदारण्यक उपनिषद्, १.३.३४.

Athaiṣa Ślōkō Bhavati। Yataścōdēti Sūryōṃstaṃ Yatra Ca Gacchatīti Pprāṇādvā Ēṣa Udēti Pprāṇēstamēti Tandēvāścakrirē Dharmmaṁ Sa Ēvādya Sa U Śca Iti Yadvā Ētēmuhryaddhriyanta Tadēvāpyadya Kurvvanti Tasmādēkamēva Vvratañcarētprāṇyāñcaivāpānyācca Nēnmā Pāpmā Mmṛtyurāpnavaditi Yadyu Carētsmāpipayiṣēttēnō Ētasyai Dēvatāyai Sāyujyaṁ Salōkatāñjayati Ya Ēvamvēda॥

Bṛhadāraṇyaka Upaniṣad, I.3.34.

32. अथ कर्म्मणाम्। आत्मेत्येतदेषामुक्थ्यमतो हि सर्व्वाणि कर्म्माण्युत्तिष्ठन्त्येतदेषाँ सामैतद्धि सर्व्वैः कर्म्माभिः सममेतदेषाम्ब्रह्मैतद्धि सर्व्वाणि कर्म्माणि बिभर्त्ति तदेतत्त्रयँ सदेकमयमात्मात्मो एकः सन्नेतत्त्रयन्तदेतदमृतँ सत्येन च्छन्नम्प्राणो वा अमृतन्नामरूपे सत्यन्ताभ्यामयम्प्राणश्छन्नः॥

बृहदारण्यक उपनिषद, I.४.३.

Atha Karmmaṇām। Ātmētyētadēṣāmukthyamatō Hi Sarvvāṇi Karmmāṇyuttiṣṭhantyētadēṣāṁ Sāmaitaddhi Sarvvaiḥ Karmmābhiḥṣamamētadēṣām Brahmaitaddhi Sarvvāṇi Karmmāṇi Bibhartti Tadētattrayaṁ Sadēkamayamātmātmō Ēkaḥ Sannētattrayantadētadamṛtaṁ Satyēna Cchannamprāṇō Vā Amṛtannāmarūpē Satyantābhyāmayamprāṇaśchannaḥ॥

Bṛhadāraṇyaka Upaniṣad, I.4.3.

33. अन्यत्र धर्मादन्यत्राधर्मादन्यत्रास्मात्कृताकृतात्।
अन्यत्र भूताच्च भव्याच्च यत्तत्पश्यसि तद्वद॥

कठोपनिषद्, १.२.१४.

Anyatra Dharmādanyatrādharmādanyatrāsmātkṛtākṛtāt।
Anyatra Bhūtācca Bhavyācca Yattatpaśyasi Tadvada॥

Kaṭhōpaniṣad, I.2.14.

34. यत्र नान्यत्पश्यति नान्यत्शृणोति नान्यद्विजानाति स भूमाथ यत्रान्यत्पश्य त्यस्यत् शृणो त्यन्यद्विजानाति तदल्पं योवै भूमा तदमृतमथ यदल्पं तन्मर्त्यं स भगवः कस्मिन् प्रतिष्ठित इति स्वे महिम्नि यदि वा न महिम्नीति।

—छांदोग्योपनिषद्, ७.२४.१.

Yatra Nānyatpaśyati Nānyatśṛṇōti Nānyadvijānāti Sa Bhūmātha Yatrānyatpaśya Tyasyat Śṛṇō Tyanyadvijānāti Tadalpaṃ Yōvai Bhūmā Tadamṛtamatha Yadalpaṃ Tanmartyaṃ Sa Bhagavaḥ Kasmin Pratiṣṭhita Iti Svē Mahimni Yadi Vā Na Mahimnīti।

Chāndōgyōpaniṣad, VII.24.1.

गोअश्वमिह महिमे त्याचक्षते हस्ति हिरण्यं दास भार्यं क्षेत्राण्यायतनानीति नाहमेवं ब्रवीमि ब्रवीमीति होवा चान्यो ह्यन्यस्मिन् प्रतिष्ठित इति॥२॥

Gōaśvamiha Mahimē Tyācakṣatē Hasti Hiraṇyaṃ Dāsa Bhāryaṃ Kṣētrāṇyāyatanānīti Nāhamēvaṃ Bravīmi Bravīmīti Hōvā Cānyō Hyanyasmin Pratiṣṭhita Iti॥2॥

35. द्वे वाव ब्रह्मणो रूपे मूर्त्तचामूर्तं चाथयन्मूर्तं तदसत्यं यदमूर्तं तत्सत्यं तद्ब्रह्म यदब्रह्म तज्ज्योतिस्स आदित्यस्सवा एष ओमि त्येतदात्मा त्रेधात्मानं व्यकुरुत ओ मिति तिस्रो मात्रा एताभिस्सर्वमिदमोतं प्रोतं चै वास्मि न्नि त्येवं ह्याहैतद्वा आदित्य ओमित्येवं ध्यायं स्तथा त्मानं युंजीतेति॥

मैत्रायणी उपनिषद्, ५.३.

Dvē Vāva Brahmaṇō Rūpē Mūrttañcāmūrtaṃ Cāthayanmūrtaṃ Tadasatyaṃ Yadamūrtaṃ Tatsatyaṃ Tadbrahma Yadbrahma Tajjyōtissa Ādityassavā Ēṣa Ōmityētadātmā Trēdhātmānaṃ Vyakuruta Ō Miti Tisrō Mātrā Ētābhissarvamidamōtaṃ Prōtaṃ Cai Vāsmi Nni Tyēvaṃ Hyāhitadvā Āditya Ōmityēvaṃ Dhyāyaṃ Stathātmānaṃ Yuñjītēti॥

Maitrāyaṇī Upaniṣad, 5.3.

सूर्यो योनिः कालस्य.....

मैत्रायणी उपनिषद्, ६.१४.

Sūryō Yōniḥ Kālasya.....

Maitrāyaṇī Upaniṣad, VI.14.

द्वे वाव ब्रह्मणो रूपे कालश्चाकालश्च अथ यः प्रागादित्यात् सोकालोकलः अथ य आदित्याद्यः स कालः सकलः।

मैत्रायणी उपनिषद्, ६.१५.

Dvē Vāva Brahmaṇō Rūpē Kālaścākālaśca Atha Yaḥ Prāgādityāt Sōkālōkalaḥ Atha Ya Ādityādyaḥ Sa Kālaḥ Sakalaḥ।

Maitrāyaṇī Upaniṣad, VI.15.

36. स होवाच। यदूर्ध्वं याज्ञवल्क्य दिवो यदवाक्पृथिव्या यदन्तरा द्यावापृथिवी इमे यद्भूतञ्च भवच्च भविष्यच्चेत्याचक्षते कस्मिन्नेव तदोतञ्च प्रोतञ्चेति॥

बृहदारण्यकोपनिषद्, ३.८.६.

Sa Hōvāca। Yadūrdhvaṃ Yājñavalkya Divō Yadavākpṛthivyā Yadantarā Dyāvāpṛthivī Imē Yadbhūtañca Bhavacca Bhaviṣya ccētyācakṣatē Kasminnēva Tadōtañca Prōtañcēti।।

Bṛhadāraṇyakōpaniṣad, III.8.6.

स होवाच। यदूर्ध्वङ्गार्गि दिवो यदवाक्पृथिव्या यदन्तरा द्यावापृथिवी इमे यद्भूतञ्च भवच्च भविष्यच्चेत्याचक्षत आकाश एव तदोतञ्च प्प्रोतञ्चेति कस्मिन्नवाकाश ओतश्च प्प्रोतश्चेति॥७॥

Sa Hōvāca। Yadūrdhvaṅgārggi Divō Yadavākpṛthivyā Yadantarā Dyāvdāpṛthivī Imē Yadbhūtañca Bhavacca Bhaviṣyaccētyācakṣata Ākāśa Ēva Tadōtañca Pprōtañcēti Kasminnavākāśa Ōtaśca Pprōtaścēti।।7।।

स होवाच एतद्वै तदक्षरङ्गार्गि ब्ब्राह्मणा अभिवदन्त्यस्थूलम् अनणवह्रस्वम् अदीर्ग्घमलोहितम् अस्नेहमच्छायमतमः अवाय्वावाकाशम् असङ्गमस्पर्शम् अगन्धमरसम् अचक्षुष्कमश्रोत्रमवागमनः अतेजस्कमप्राणममुखमनामागोत्रम् अजरममरमभयम् अमृतमरजः अशब्दमविवृतमसम्वृतमपूर्व्वम् अनपरमनन्तरम-बाह्यम् न तदश्नोति कञ्चन॥८॥

Sa Hōvāca। Ētadvai Tadakṣaraṅgārggi Bbrāhmaṇā Abhivadantya-sthūlam Anaṇavahrasvam Adīrggham Alōhitam Asnēham Acchāyamatamaḥ Avāyvāvākāśam Asaṅgam Asparśam Agandham Arasam Acakṣuṣkam Aśrōtramavāgamanaḥ Atējaskamapprāṇam Amukham Anāmāgōtram Ajaramamaramabhayam Amṛtamarajaḥ Aśabdamavivṛtamasamvṛtamapūrvvam Anaparamanantarama-bāhyam Na Tadaśnōti Kañcana।।8।।

एतस्य वा अक्षरस्य प्प्रशासने गार्गि द्यावापृथिवी व्विधृते तिष्ठतः एतस्य वा अक्षरस्य प्प्रशासने गार्गि सूर्याचन्द्रमसौ व्विधृतौ तिष्ठतः एतस्य वा अक्षरस्य प्प्रशासने गार्ग्यहोरात्राण्यर्द्धमासा मासा ऋतवः सम्वत्सरा व्विधृतास्तिष्ठन्त्येतस्य वा अक्षरस्य प्प्रशासने गार्गि प्प्राच्यो अन्या नद्यः स्यन्दन्ते श्वेतेभ्यः पर्व्वतेभ्यः प्प्रतीच्यो अन्या यां याञ्च दिशमेतस्य वा अक्षरस्य प्प्रशासने गार्गि ददतम्मनुष्याः प्प्रशँसन्ति यजमानन्देवा दर्व्वीम्पितरोन्वायत्ताः॥९॥

Ētasya Vā Akṣarasya Ppraśāsanē Gārggi Dyāvāpṛthivī Vvidhṛtē Tiṣṭhataḥ Ētasya Vā Akṣarasya Ppraśāsanē Gārggi Sūryācandra-

masau Widhṛtau tiṣṭhataḥ Ētsbya vā Akṣarasya Ppraśāsanē Gārggyahōrātrāṇyard-dhamāsā Māsā Ṛtavaḥ Samvatsarā Vvidhṛtāstiṣṭhantyētasya Vā Akṣarasya Ppraśāsanē Gārggi Pprācyō Anyā Nadyaḥ Syandantē Śvētēbhyaḥ Parvvatēbhyaḥ Ppratīcyō Anyā Yāṃ Yāñca Diśamētasya Vā Akṣarasya Ppraśāsanē Gārggi Dadatammanuṣyāḥ Ppraśaṁsanti Yajamānandēvā Darvvyampitarōnvāyattāḥ//9//

यो वा एतदक्षरमविदित्वा गार्ग्गी। अस्मिंल्लोके जुहोति ददाति तपस्यत्यापि बहूनि व्वर्षसहस्राण्यन्तवानेवास्य स लोको भवति यो वा एतदक्षरमविदित्वा गार्ग्यस्मा- ल्लोकात्पैति स कृपणोथ य एतदक्षरङ्गार्ग्गी व्विदित्वास्माल्लोकात्प्रैति स ब्ब्राह्मणः॥१०॥

Yō Vā Ētadakṣaramaviditvā Gārggī/ Asmiṃllōkē Juhōti Dadāti Tapasyatyāpi Bahūni Vvarṣasahasrāṇyantavānēvāsya Sa Lōkō Bhavati Yō Vā Ētadakṣaramaviditvā Gārggyasmāllōkātpaiti Sa Kṛpaṇōtha Ya Ētadakṣaraṅgārggī Vviditvāsmāllōkātpraiti Sa Bbrāhmaṇaḥ//10//

37. कालः स्वभावो नियतिर्यदृच्छा भूतानि योनिः पुरुष इति चिन्त्या।
संयोग एषां न त्वात्मभावादात्माप्यनीशः सुखदुःखहेतोः॥

श्वेताश्वतरोपनिषद्, १.२.

Kālaḥ Svabhāvō Niyatiryadṛcchā Bhūtāni Yōniḥ Puruṣa iti Cintyā। Saṃyōga Ēṣāṃ Na Tvātmabhāvādātmāpyanīśaḥ Sukhaduḥkhahētōḥ// Śvētāśvatarōpaniṣad, 1.2.

38. द्वे वाव ब्ब्रह्मणो रूपे। मूर्त्तञ्चैवामूर्तञ्च मर्त्त्यञ्चामृतञ्च स्थितञ्च यच्च सच्च त्यच्च॥ बृहदारण्यकोपनिषद्, २.३.१.

Dvē Vāva Bbrahmaṇō Rūpē/ Mūrttañcaivāmūrtañca Marttyañcāmṛtañca Sthitañca Yacca Sacca Tyacca//

Bṛhadāraṇyakōpaniṣad, II.3.1.

तदेतनमूर्त्तम्। यदन्यद्वायोश्चान्तरिक्षाच्चैतन्मर्त्त्यमेतत्स्थितमेतत्सत॥२॥

Tadētanmūrttam। Yadanyatvāyōścāntarikṣāccaitanmarttyam Ētatsthita Mētatsata//2//

तस्यैतस्य मूर्त्तस्य। एतस्य मर्त्त्यस्यैतस्य स्थितस्यैतस्य सत एष रसो य एष तपति सतो ह्येष रसः॥३॥

Tasyaitasya Mūrttasya/ Ētasya Marttyasyaitasya Sthitasyaitasya Sata Ēṣa Rasō Ya Ēṣa Tapati Satō Hyēṣa Rasaḥ//3//

अथामूर्त्तम्। व्वायुश्चान्तरिक्षञ्चैतदमृतमेतद्यदेतत्त्यम॥४॥

Athāmūrttam/ Vvāyuścāntarikṣañcaitadamṛtamētadyadētatt yama//4//

तस्यैतस्यामूर्त्तस्य। एतस्यामृतस्यैतस्य यत एतस्य त्यस्यैष रसो य एष एतस्मिन्मण्डले पुरुषस्त्यस्य ह्येष रस इत्याधिदेवतम्॥५॥

Tasyaitasyāmūrttasya । Ētasyāmṛtasyaitasya Yata Ētasya Tyasyaiṣ a Rasō Ya Ēṣa Ētasminmaṇḍalē Puruṣastyasya Hyēṣa Rasa Ityādhidēvatam॥5॥

अथाध्यात्मम्। इदमेव मूर्त्तं यदन्यत्प्राणाच्च यश्चायमन्तरात्मन्नाकाश एतन्मर्त्त्य मेतत्स्थित मेतत्सत्॥६॥

Athādhyātmam। Idamēva Mūrttaṃ Yadanyatprāṇācca Yaścāya-mantarātman Nākāśa Ētanmarttyamētatsthitamētatsat॥6॥

तस्यैतस्य मूर्त्तस्य। एतस्या मर्त्त्यस्यैतस्य स्थितस्यैतस्य सत एष रसो यच्चक्षुः सतो ह्येष रसः॥७॥

Tasyaitasya Mūrttasya। Ētasyā Marttyasyaitasya Sthitasyaitasya Sata Ēṣa Rasō Yaccakṣuḥ Satō Hyēṣa Rasaḥ॥7॥

अथामूर्त्तम्। प्राणश्च यश्चायमन्तरात्मन्नाकाशः एतदमृतमेतद्यदेतत्त्यम्॥८॥

Athāmūrttam। Pprāṇaśca Yaścāyamantarātmannākāśaḥ Ētadamṛta-mētadyadētattyam॥8॥

तस्यैतस्यामूर्त्तस्य। एतस्यामृतस्यैतस्य यत एतस्य त्यस्यैष रसो योयन्दक्षिणेक्षन्पुरुष-स्त्यस्य ह्येष रसः॥९॥

Tasyaitasyāmūrttasy। Ētasyāmṛtasyaitasya Yata Ētasya Tyasyaiṣa Rasô Yāyandakṣiṇēkṣanpuruṣastyasya Hyēṣa Rasaḥ॥9॥

तस्य हैतस्य पुरुषस्य रूपम्। यथा माहारजनम्बासो यथापाण्ड्वाविकं यथेन्द्रगोपो यथाग्न्यर्च्चिर्य्यथा पुण्डरीकं यथा सकृद्विद्युत्तँ सकृद्विद्युत्तेव ह वा अस्य श्रीर्भ्भवति य एवम्वेद॥१०॥

Tasya Haitasya Puruṣasya Rūpam। Yathā Māhārajanamvāsō Yathā-pāṇḍvāvikaṃ Yathēndragōpō Yathāgnyarcciryyathā Puṇḍarīkaṃ Yathā Sakṛdvidyuttaṁ Sakṛdvidyuttēva Ha Vā Asya Śrīrbhbhavati Ya Ēvamvēda॥10॥

अथात आदेशः। नेति नेति न ह्येतस्मादिति नेत्यन्य त्परमस्त्यथ नामधेयँ सत्त्यस्य सत्त्यमिति प्राणा वै सत्त्यन्तेषामेषसत्त्यम्॥११॥

Athāta Ādēśaḥ। Nēti Nēti Na Hyētasmāditi Nētyanya Tparamast-yatha Nāma Dhēyaṁ Sattyasya Sattyamitti Pprāṇā Vai Sattyantēṣ āmēṣasattyam॥11॥

39. Maitri Upaniṣad, VII, II, 8.

40. स होवाच बालाकि र्य एवैष आदित्ये पुरुषस्तमेवाह मुपास इति तं होवाचा

जातशत्रुर्मा मैतस्मिन् समवादयिष्ठा बृहन् पांडरवासाधिष्ठा सर्वेषां भूतानां मूर्थेतिवा अहमेत मुपास इति सयोहैत मेव मुपास्तेधिष्ठा सर्वेषां भूतानां मूर्धाभवति॥

कौषीतकी उपनिषद्, ४.२.

Sahōvāca Bālāki Rya Ēvaiṣa Ādityē Puruṣastamēvāha Mupāsa Iti Taṃ Hōvācā jātaśatrurmā Maitasmin Samavādayiṣṭhā Bṛhan Pāṇḍaravāsādhiṣṭhā Sarvēṣāṃ Bhūtānāṃ Mūrthētivā Ahamēta Mupāsa Iti Sayōhaita Mēva Mupāstēdhiṣṭhā Ssarvēṣāṃ Bhūtānāṃ Mūrdhābhavati//

Kauṣītakī Upaniṣad, IV. 2.

41. तस्यैष आदेशो य देतद् विद्युतोव्यद्युतदास इतीति न्यमीमिषदास इत्यधिदैवतम्॥

केनोपनिषद्, ४.४.

Tasyaiṣa Ādēśō Ya Dētad Vidyutōvyadyutadāsa Itīti Nyamīmiṣadāsa Ityadhi Daivatam॥

Kēnōpaniṣad, IV.4.

अथाध्यात्मं य देतद् गच्छतिव च मनोनेन चैत दुपस्मर त्यभीक्ष्णं संकल्पः॥५॥

Athādhyātmaṃ Ya Dētad Gacchativa Ca Manōnēna Caita Dupasmara Tyabhīkṣṇaṃ Saṅkalpaḥ//5//

42. सा ब्रह्मेति होवाच ब्रह्मणो वा एतद्विजये।
महीयध्व मिति ततो हैव विदांचकार ब्रह्मेति॥

केनोपनिषद्, ४.१.

Sā Brahmēti Hōvāca Brahmaṇō Vā Ētadvijayē।
Mahīyadhva Miti Tatō Haiva Vidāñcakāra Brahmēti//

Kēnōpaniṣad, IV.1.

तस्माद् वा इंद्रोति तरामि वान्यान् देवान् स ह्येनन्ने।
दिष्ठं पस्पर्श स ह्येनत् प्रथमो विदांचकार ब्रह्मेति॥३॥

Tasmād Vā Iṃdrōti Tarāmi Vānyān Dēvān Sa Hyēnannē।
Diṣṭhaṃ Pasparśa Sa Hyēnat Prathamō Vidāñcakāra Brahmēti//3//

43. ब्रह्मणो वावैता अध्यास्तनवः परस्यामृतस्याशरीरस्य।

मैत्रायणी उपनिषद्, ४.६.

Brahmaṇō Vāvaitā Adhyāstanavaḥ Parasyāmṛtasyāśarīrasya/

Maitrāyaṇī Upaniṣad, IV.6.

44. यः कालं ब्रह्मेत्युपासीत कालस्तस्यातिदूरमपसरति।

मैत्रायणी उपनिषद्, ६.१४.

Yaḥ kālaṃ Brahmētyupāsīta Kālastasyātidūramapasarati/

Maitrāyaṇī Upaniṣad, VI.14.

45. Svāmi Mādhava Tīrtha, The Concept of Time in Indian Philosophy, p. 1-2.

46. Maitrāyaṇī Upaniṣad, VI.14.

47. स्वभावमेके कवयो वदन्ति कालं तथान्ये परिमुह्यमानाः।
देवस्यैष महिमा तु लोके येनेदं भ्राम्यते ब्रह्मचक्रम्॥ श्वेताश्वतरोपनिषद्, ६.१.
Svabhāvamēkē Kavayō Vadanti Kālaṃ Tathānyē Parimuhyamānāḥ I
Dēvasyaiṣa Mahimā Tu Lōkē Yēnēdaṃ Bhrāmyatē Brahmacakram II
Śvētāśvatarōpaniṣad, VI.1.

48. येनावृतं नित्यमिदं हि सर्वं ज्ञः कालकारो गुणी सर्वविद्यः।
तेनेशितं कर्म विवर्तते ह पृथ्व्यप्तेजोनिलखानि चिन्त्यम्॥
श्वेताश्वतरोपनिषद्, ६.२.
Yēnāvṛtaṃ Nityamidaṃ Hi Sarvaṃ Jñaḥ Kālakārō Guṇī Sarvavidyaḥ I
Tēnēśitaṃ Karma Vivartatē Ha Pṛthvyaptējōnilakhāni Cintyam II
Śvētāśvatarōpaniṣad, VI.2.
तत्कर्म कृत्वा विनिवर्त्य भूयस्तत्त्वस्य तत्त्वेन समेत्य योगम्।
एकेन द्वाभ्यां त्रिभिरष्टभिर्वा कालेन चैवात्मगुणैश्च सूक्ष्मैः॥३॥
Tatkarma Kṛtvā Vinivartya Bhūyastattvasya Tattvēna Samētya Yōgam I
Ēkēna Dvābhyāṃ Tribhiraṣṭabhirvā kālēna Caivātmaguṇaiśca Sūkṣmaiḥ II3II

49. आदिः स संयोगनिमित्तहेतुः परस्त्रिकालादकलोपि दृष्टः।
तं विश्वरूपं भवभूतमीड्यं देवं खचित्तस्थमुपास्य पूर्वम्॥
श्वेताश्वतरोपनिषद्, ६.५.
Ādiḥ Sa Saṃyōganimittahētuḥ Parastrikālādakalōpi Dṛṣṭaḥ I
Taṃ Viśvarūpaṃ Bhavabhūtamīḍyaṃ Dēvaṃ Khacittasthamupāsya Pūrvam II
Śvētāśvatarōpaniṣad, VI.5.
स वृक्षकालाकृतिभिः परोन्यो यस्मात्प्रपञ्चः परिवर्ततेयम्।
धर्मावहं पापनुदं भगेशं ज्ञात्वात्मस्थममृतं विश्वधाम॥६॥
Sa Vṛkṣakālākṛtibhiḥ Parōnyō Yasmātprapañcaḥ Parivartatēyam I
Dharmāvahaṃ Pāpanudaṃ Bhagēśaṃ Jñātvātmasthamamṛtaṃ Viśvadhāma II6II

50. येनावृतं नित्यमिदं हि सर्वं ज्ञः कालकारो गुणी सर्वविद्यः।
तेनेशितं कर्म विवर्तते ह पृथ्व्यप्तेजोनिलखानि चिन्त्यम्॥
श्वेताश्वतरोपनिषद्, ६.२.

Yēnāvṛtaṃ Nityamidaṃ Hi Sarvaṃ Jñaḥ Kālakārō Guṇī Sarvavidyaḥ।
Tēnēśitaṃ Karma Vivartatē Ha Pṛthvyaptējōnilakhāni Cintyam॥
Śvētāśvatarōpaniṣad, VI.2.

51. तुसाभक्षं तु मां ज्ञात्वा प्रविवित्तजने गृहे।
बिभ्रतां गार्दभं रूपमागत्य परिगर्हसे॥

महाभारतम्, १२.२३१.२५.

Tusābhakṣaṃ Tu Māṃ Jñātvā Pravivittajanē Gṛhē।
Bibhratāṃ Gārdabhaṃ Rūpamāgatya Parigarhasē॥
Mahābhārata, XII.231.25.

तत्र चेन्न भवेदेवं संशयः कर्मनिश्चये।
किम्नु कर्मस्वभावोयं ज्ञानं कर्मेति व पुनः॥

महाभारतम्, १२.२४४.२.

Tatra Cēnna Bhavēdēvaṃ Saṃśayaḥ Karmaniścayē।
Kimnu Karmasvabhāvōyaṃ Jñānaṃ Karmēti Va Punaḥ॥
Mahābhārata, XII.244.2.

52. अस्यः यतिः कूतो वेति ब्राह्मणः श्रोत्रियः परः।
विद्वान्विप्रो ह्यकुटुम्बः प्रियवादि महानापः॥

महाभारतम्, १२.२२४.५.

Asyaḥ Yatiḥ Kūtō Vēti Brāhmaṇaḥ Śrōtriyaḥ Paraḥ।
Vidvānviprō Hyakuṭumbaḥ Priyavādi Mahānāpaḥ॥
Mahābhārata, XII.244.5.

त्वमप्याहमहं सर्वमहमित्येव वर्तते।
नत्र तत्परमार्थं वै किमर्थमनुपृच्छसि॥ महाभारतम्, १२.२२४.५६.
Tvamapyāhamahaṃ Sarvamahamityēva Vartatē।
Natra Tatparamārthaṃ Vai Kimarthamanupṛcchasi॥
Mahābhārata, XII.224.56.

53. *Ibid.*, 227.73.
54. *Ibid.*, 227, 81-2.
55. *Ibid.*, 222, 85.
56. *Ibid.*, 227, 103.
57. *Ibid.*, 231, 23 & 27.
58. *Ibid.*, 27, 44.
59. *Ibid.*, 224.
60. Raimundo Panikkar, 'Time and History in the Tradition of India : Kāla and Karma', In Culture and Time, Ed., L. Gaslet et al, Paris, UNESCO Press, 1976, p. 26-27.

61. स्वभावादेव संदृश्य वर्तमानाः प्रवृत्तयः।
स्वभावनिरताः सर्वाः परितप्ये न केनचित्॥

महाभारतम्, १२.१७२.११.

Svabhāvādēva Sandṛśya Vartamānāḥ Pravṛttayaḥ I
Svabhāvaniratāḥ Sarvāḥ Paritapyē Na Kēnacit II

Mahābhārata, XII.172.11.

62. पश्यन् प्रह्राद भूतानामुत्पत्तिमनिमित्ततः।
ह्रासं वृद्धिं विनाशं च न प्रहृष्ये न च व्यथे॥

महाभारतम्, १२.१७२.१०.

Paśyan Prahrāda Bhūtānāmutpattimanimittataḥ I
Hrāsaṃ Vṛddhiṃ Vināśaṃ Ca Na Prahṛṣyē Na Ca Vyathē II

Mahābhārata, XII.172.10.

63. तृषितमनियतं मनो नियन्तुं व्रतमिदमाजगरं शुचिश्चरामि।

महाभारतम्, १२.१७२.३३.

Tṛṣitamaniyataṃ Manō Niyantuṃ Vratamidamājagaraṃ Śuciśca-rāmi II Mahābhārata, XII.172.33.

64. एकः शास्ता न द्वितीयोस्ति शास्ता गर्भे शयानं पुरुषं शास्ति शास्ता।
तेनानुशिष्टः प्रवणादिवोदकं यथा नियुक्तोस्मि तथा वहामि॥

महाभारतम्, १२.२१९.८.

Ēkaḥ Śāstā Na Dvitīyōsti Śāstā Garbhē Śayānaṃ Puruṣaṃ Śāsti Śāstā I
Tēnānuśiṣṭaḥ Pravaṇādivōdakaṃ Yathā Niyuktōsmi Tathā Vahāmi II

Mahābhārata, XII.219.8.

65. केचित् पुरुषकारं तु प्राहुः कर्मविदो जनाः।
दैवमित्यपरे विप्राः स्वभावं भूतचिन्तकाः॥

महाभारतम्, १२.२२४.५०.

Kēcit Puruṣakāraṃ Tu Prāhuḥ Karmavidō Janāḥ I
Daivamityaparē Viprāḥ Svabhāvaṃ Bhūtacintakāḥ II

Mahābhārata, XII.224.50.

येभ्यः सृजति भूतानि कालो भावप्रचोदितः।
महाभूतानि पञ्चेति तान्याहुर्भूतचिन्तकाः॥

महाभारतम्, १२.२६७.४.

Yēbhyaḥ Sṛjati Bhūtāni Kālō Bhāvapracōditaḥ I
Mahābhūtāni Pañcēti Tānyāhurbhūtacintakāḥ II

Mahābhārata, XII.267.4.

विद्धि नारद पञ्चैतान् कालषष्ठान् स्वभावतः॥ महाभारतम्, १२.२६७.६.
Viddhi Nārada Pañcaitān Kālaṣaṣṭhān Svabhāvataḥ॥
Mahābhārata, XII.267.6.

66. कालोस्मि लोकक्षयकृत्प्रवृद्धो लोकान् समाहर्तुमिह प्रवृत्तः।
ऋतेपि त्वा न भविष्यन्ति सर्वे ये वस्थिताः प्रत्यनीकेषु योधाः॥
भगवद्गीत, ११.३२.
Kālōsmi Lōkakṣayakṛtpravṛddhō Lokān Samāhartumiha Pravṛttaḥ।
Ṛtēpi Tvā Na Bhaviṣyanti Sarvē Yē Vasthitāḥ Pratyanīkēṣu Yōdhāḥ॥
Bhagavadgīta, XI.32.

67. लेलिह्यसे ग्रसमानस्समन्ता ल्लोकान् समग्रान् वदनैर्ज्वलद्भिः।
तेजोभिरापूर्य जगत्पमग्रं भासस्तवोग्राः प्रतपन्ति विष्णो॥ भगवद्गीत, ११.३०.
Lēlihyasē Grasamānassamantā Llōkān Samagrān Vadanairjvaiad-bhiḥ।
Tējōbhirāpūrya Jagatpamagraṃ Bhāsastavōgrāḥ Pratapanti Viṣṇō॥ Bhagavadgīta, XI.30.

68. *Ibid.*, IV.10-20.

69. अनादिर्भभगवान्कालो नान्तोस्य द्विज विद्यते॥
अव्युच्छिन्नास्ततस्त्वेते सर्गस्थित्यन्तसंयमाः॥ विष्णु पुराणम्, १.२.२६.
Anādirbhabhagavānkālō Nāntōsya Dvija Vidyatē॥
Avyucchinnāstatastvētē Sargasthityantasaṃyamāḥ॥
Viṣṇu Purāṇa, 1.2.26.

70. Bhāgavata Purāṇa, VI.11.19.
71. *Ibid.*, II.26.18.
72. Bhāgavata Purāṇa, III.8.11.
73. *Ibid.*, 1.2.
74. *Ibid.*, XI.6.9; 8.28; 9.14; 13. 16.
75. Sadananda Bhaduri, 'Time (in Nyāya-Vaiśēsika)' studies in Nyāya-Vaiśēṣika Metaphysics, Pune, Bhandarkar Oriental Research Institute, 1975, p. 183.
76. A.B. Keith, 'Time and Space (in Nyāya-Vaiśēṣika)', Indian Logic and Atomism : An Exposition of the Nyāya and Vaiśēṣika System, Delhi, Oriental Books Reprint Corporation, 1977, p. 233.
77. *Ibid.*
78. *Ibid.*
79. *Ibid.*, 234.
80. Indukula Jhaveri, 'Concept of Kāla and Ākāśa in the Sāṅkhya-Yōga System', Journal of Oriental Institute (Baroda) 5, 1955, p. 1.
81. *Ibid.*, p. 2.
82. *Ibid.*, pp. 2-3.

CHAPTER - II

THE CONCEPT OF TIME IN JAINISM

In the post-Vēdic times ancient India witnessed the rise of two great religions—*Jainism* and *Buddhism*. Jainism is the older religion of the two. According to the Jain literature there have been 24 *Tīrthaṅkaras* as spiritual *Gurus* of the Jains. The first of them was *Ṛṣabha* and the last one *Mahāvīra,* is senior contemporary of *Buddha*.

The Jain philosophy does not believe in the creation of the world by God or one Divine Being. There must be at least two, Matter and Spirit. This is something like the *Sāṅkhya* philosophy. The only difference between the two is that in *Sāṅkhya, Puruṣa* undergoes no change internal or external, whereas in the Jain philosophy *Ātma* or soul undergoes *Svapariṇāma,* i.e., internal change. Besides, Time and Space are taken as fundamentals in the Jain philosophy although in one sense they are relative also.

In the Jaina philosophy, *Kāla* is as much a real substance as the five others,[1] viz., *Jīva, Dharma, Adharma, Pudgala* and *Ākāśa.* It is described as the accompanying cause or condition *(Sahakārī Kāraṇa)* or *Asamavāyī Kāraṇa,* as the *Vaiśēṣikas* call it, of the modification *(Pariṇāma)* of substances.[2] Substances are modified because of their own nature. Yet, Time is the real factor in the happening of their modifications.[3] As the modifications *(Pariṇāma)* of substances are due to their own nature, one need not admit the real existence of time, as their accompanying cause. Substances move, come to rest and have spatial existence, of and by themselves; yet the Jaina Scripture posits the reality of *Dharma, Adharma* and *Ākāśa* as the accompanying condition of the motion, the rest and the spatial existence of substances. Similarly, Time should be admitted as the accompanying condition of the modifications of substances.[4] Indeed, the necessity of admitting the real existence is stranger in the case of Time. We come across moments, hours, days etc., in our actual observation, these are the effects or limitations which point to the reality of a principle which we call *Kāla*. It is thus, that observation confirms our belief in the reality of Time. The opponents of the Jaina theory of Time contend that Time is no reality. They point out that in the *Jaina Āgama 'Samaya'* (a duration or measure of time) is described as the time taken by an atom in crossing over a *Pradēśa* of *Ākāśa* and that the same *Āgama* elsewhere talks of the crossing of the whole universe in the course of one *Samaya*. This shows that Time is no reality; it is more or less a convention. The Jaina commentator, *Brahmadēva,* sets aside this contention by pointing out that the above different statements in the *Āgama* refer to the differences in speed and do not affect the reality of Time. "Dēvadatta by moving slowly", says he, 'traverses a distance of 100 Yōjanas in one hundred days; he may, however, acquire super-human powers and swiftly pass over the same distance in one day. The Time is real notwithstanding the different results effected by difference in speed."

The next objection is that there need not be supposed any real Time to explain motion and change in things. *Dharma,* which is admitted by the Jainas as a real substance, is what accounts for motion; why should *Kāla* be admitted over and above Dharma? The Jainas answer that the accompanying conditions of a phenomenon are often many in number. An animal, for instance, moves of itself; yet *Karma* is postulated to determine its motion. And not duly that, but in the case of fishes, for example, water is admitted as an accompanying condition of motion besides *Karma.* This shows that there may be more than one accompanying condition. Hence, there is no inconsistency if *Kāla* be supposed to be a reality and accompanying condition of motion and change.

Then, again there may be objectors who may admit the reality of Time but find difficulty in looking upon it as a condition of the modification of substances. They point out that according to the Jainas themselves, Time is confined within the limits of *Lōkākāśa* or "the filled space" only; the *Alōka* which is beyond this has no Time within it. But the *Alōka* is a substance all the same and must have modifications. How are these modifications possible in the *Alōka* which has no Time within it? The Jaina thinkers answer this objection by stating that the *Alōka,* although it is beyond the *Lōka* is still a part of the *Ākāśa* or space; Time being within the *Ākāśa,* it brings about modifications in any part of it, just as a potter's stick moves the whole wheel by striking at a particular point of it and just as a pleasant object coming in contact with a particular part causes pleasant feeling all over the body.

In Jaina philosophy a real thing is said to be characterized by its three aspects, respectively called *Utpāda* or origination, *Vyaya* or annihilation, and *Dravya* or persistence. In other words, whatever is real considered to come into manifestation to enter into annihilation, so far as a particular manifestation or modification is concerned and to persist, so far as its essential substance is concerned. The same fact about a real thing is otherwise stated by the Jaina philosophers by recognising in it two aspects, the *Paryāya* or the mode and the *Dravya* or the substance. The former is the series of temporary modes with come *(Utpāda)* and go *(Vyaya)* and the latter is the essentiality which is constant *(Dhrauvya)*.

Time, as a real substance, has three aspects of the *Utpāda*. The *Vyaya* and the *Dhrauvya,* and its two aspects of the *Paryāya* and the *Dravya* may also be distinguished.

Let us suppose the phenomena of a man's clenching his fingers into a fist. When the clenching of the fingers occurs *(Utpāda),* the previous state of the fingers necessarily is at an end *(Vināśa* or *Vyaya),* yet, so far as the fingers are concerned, they continue to be substantially the same *(Dhrauvya)*. Or again, when pure knowledge arises *(Utpāda)* as on effect, its cause, and disturbed contemplation, is at an end *(Vyaya);* and yet, the same soul in its pure substance underlies both the phenomena *(Dhrauvya)*. In the same way, a certain material phenomenon indicates the time which we call, the present; when this present time arises *(Utpāda),* the time which preceded it is at an end *(Vyaya);* and we shall see presently, how the noumenal time underlies both these phenomena of time *(Dhrauya)*. Thus, it is that the three aspects of Time—its origination, its annihilation and its persistence—may be easily distinguished and that accordingly Time is a substance.

Time in its impermanent aspects, viz., the aspects of origination and decay—*Utpāda* and *Vyaya*—is called the *Paryāya Kāla* (Time in modification) or *Vyavahāra Kāla*[5] (Time in ordinary practice). It is technically called the *Samaya*. What we call moments, hours, days, etc., are instances of *Samaya* or Empirical Time, as these have both beginning and end. The author of the *Tattvārthasāra* describes the *Vyavahāra Kāla* in the following way; "The great sages say that the practical time is indicated by modification, activity, distance and proximity". Modification is described by the Jains as the self-variation in a thing without a change in its substance. Activity is due to the movement, etc., of a body. Distance consists in being far away and proximity, in being near; these two, viz., distance and proximity are to be understood here in relation to Time. In other words, we call a thing new or old (in time) by looking at its state of modification, on seeing the rise of the Sun in the East, we say we have the day-break (morning time) and so our ideas of mid-day, evening, night etc., are determined by our observation of the movements of heavenly bodies. A thing under actual observation is said to be proximate in time, otherwise it is distant in time. In this way, the outside objects and phenomena indicate and refer to the *Vyavahāra Kāla* or *Samaya*,[6] i.e., temporary time denoted by moments, days, hours etc., as we call them.

To say that Samaya is indicated by the outside phenomena is, however, very difficult from saying that *Kāla* or Time is nothing real but is only the effect of those outside material phenomena. There is a class of thinkers who maintain that the Time which we call *Nimēṣa* is but an effect produced by the movements of our eye-lids. Similarly, an hour, according to them, is but the effect of things like the water-clock, etc., and a day, of the movement of the Sun. Aristotle, for example, by closely connecting motion and time, was almost an upholder of this theory. The effect of material things would bear the stamp of materiality. Food, for instance, which is the effect of rice (boiled) has colour (black, white, etc.), smell (good, pleasant, etc.), touch (soft, hard, etc.) and taste (sweet, etc.). If time were nothing but an effect of material things or phenomena, it would have had colour and other attributes of matter. Hence, it is that Time must be supposed to be a reality, distinct from matter.[7]

Paryāya or modifications refer to *dravya* or the substance which persists in those modifications. Hours, minutes, days, etc., are periods which have a beginning and an end. These point to the Time which continues to persist. This is Time in its *Dhrauvya* aspect, *Dravya Kāla* or *Niścaya Kāla,* as it is technically called. Some thinkers contend that there is no (Noumenal) *Kāla* besides the *Samaya,* in as much as there is no evidence of a *Real Kāla.* But it may be pointed out that *Samaya* as shown above, has a beginning and an end. It is accordingly a *Paryāya,* an evanescent state which is impossible without a *dravya* or persisting substance behind it, as its material cause and support. Food, for instance is a *Paryāya,* an effect; although fire, fuel, etc., are operating conditions towards its production, it points to rice as the *dravya* or the substantial cause. A potter, his wheel etc., are no doubt necessary for the production of an earthern pitcher, an effect or *Paryāya;* yet, it refers to clay as its *dravya* or material cause. Lastly, one is a human being, sometimes; he remains in hell or heaven, sometimes; these states suppose some substance, i.e., the soul, which persists in these states. It is thus that any phenomenon which is a *Paryāya* always refers to a *dravya* which underlies it and out of which it arises and into which it disappears. *Samaya* thus proves *Kāla,* a Noumenal Time.

What is this *Niścaya* or *Dravya Kāla?* The Jaina philosophers say that *Paramārtha Kāla* or Noumenal Time is characterized by *Vartana. Vartana* consists in an apprehension of continuity of a substance. A material thing, for instance undergoes changes from moment to moment. Observation shows that besides the consciousness of a material substance underlying those changes, there is another consciousness of a continuity of something real throughout those moments of change. This latter consciousness of continuity is *Vartana* and Noumenal Time is characterised by it. The commentator *Brahmadēva* thus explains *Vartana* and *Dravya Kāla*—"Things change of and by themselves and their own essential nature. *Vartana* is the accompanying cause of the modification of those things—like the basal stone in a potter's wheel, like fire in the matter of studying in winter-time. This *Vartana* is the characteristic of Real Time." In other words, Real Time, although it does not cause the changes in things is nevertheless an invariable accompanying condition of these. The stone underneath the potter's wheel doesn't cause motion to the wheel; but in the matter of the movement of the wheel, the stone is indispensable. Fire, again, doesn't cause one's study in winter; but study is impossible without fire. So is the case with time. The position thus ultimately comes to this, that changes in things point to something more than those changes and the changing things. Every change is accompanied by a temporary period or point of time, called *Samaya* or Empirical Time, while the continuing reality underlying the Empirical Time is the *Dravya Kāla*[8] or Noumenal Time.

Time as a substance is obviously similar to the soul but in as much as it is essentially unconscious, it is distinct from the soul and similar to the other *Ajīvas*. Time is *Amūrta* or formless; in this respect, it is similar not only to the soul but to the non-souls except matter. *Kāla* is described as *Niṣkriya*, i.e., devoid of activity and is different from soul and matter which are *Sakriya* or active; other *Niṣkriya* substances are *Dharma, Adharma* and *Ākāśa*.

The most important thing to be noticed in this connection is the classification of substances into extended and non-extended. *Jīva, Pudgala, Ākāśa, Dharma* and *Adharma* are called *Astikāyas* or extended substances, while *Kāla* is non-extended. An extended substance is that which has many *Pradēśas*. A *Pradēśa* is that portion of space which is obstructed by one ultimate unit or atom of matter. *Ākāśa* or Space is infinite and hence its *Pradēśas* are 'infinite'. *Dharma* and *Adharma* pervade the *Lōkākāśa* but they do not extend beyond it. Accordingly, the *Pradēśas* of *Dharma* and *Adharma* are said to be 'innumerable' (and not 'infinite') in as much as they have a limit. A *Jīva* also is capable of filling up measurable or indefinite (though not infinite) parts of space and its *Pradēśas* are also 'innumerable'.

According to some, there is no other time except that which consists of acts comprised by the rising and setting of the Sun.[9]

As regards matter, it may be said that the parts of space, occupied by one or two molecules can be counted; in such cases, matter may be said to have 'innumerable' *Pradēśas,* such material molecules fill the *Lōkākāśa*[10] and do not extend beyond it and hence matter may also be said to have 'innumerable' *Pradēśas*. And thirdly, atoms of matter in their subtle state are infinite and accordingly, matter may be said to have an 'infinite' number of *Pradēśas*.

A minute part of a substance is called *Aṇu* and when these *Anus* are combined inseparably, the substance constituted of them is an *Astikāya* or extended substance. *Jīva, Pudgala, Ākāśa, Dharma* and *Adharma* are extended substances, because their minutest constituents are mixed up and inseparably combined with one another. Such is, however, not the case with Time. It has no doubt, its minute parts, the *Kālāṇus*[11] or instants; but each of these *Kālāṇus* is strictly separate from the other; the minute parts of Time are never mixed up with one another. This is the reason why the Jaina writers compare Time-substance with a heap of jewels. Time is not an *Astikāya* and in as much as each of its minute part is strictly individual, it may be said to have only one *Pradēśa,* or rather, no *Pradēśa* at all.

If the minute points of Time are indefinitely many in number, filling, as they do, the whole of the *Lōkākāśas,* it is clear that the Jaina view is opposed to other Indian theories according to which Time is one. The Jaina thinkers maintain that if Time were one, there would have been no distinction between past, present and further phenomena, all phenomena would have been perceived simultaneously in one unvaried 'now'. It is only when we regard the time-units as separate and many in number that we can correctly explain our perception of phenomena which are simultaneous or of long duration or of short duration. The phenomena which occur in one point of time are simultaneous; the phenomena which occur in many points of time are of long duration and those which engage fewer points of time are of short duration. The objections contend that Time is one and that its variety as past, present, future etc., are due to differences in *Upādhis* or limitations.

It is said that Time is one and that because the phenomena occur in varied orders, Time appears to us to be varied as present, past and future. The Jainas point out that the varied order of their phenomena indicates nothing but variation in time-units. The Jainas hold that time-units are indefinitely many in number and that they fill every minutest point in filled space *(Lōkākāśa)*. It is said that as Time which is existent in *Lōkākāśa* can account for changes in the *Alōka,* it may be one and yet effects changes in every part of the *Lōkākāśa.* The Jainas point out that the contention is not at all sound. Time can effect changes in the *Alōka* because the *Alōka* is a part of the *Ākāśa* which is but whole. In *Lōkākāśa,* however, *pudgalas* and souls are infinitely many in number, their *Pradēśas* occupying its various parts so that Time which effects changes in *pudgalas* and souls must consequently be many in number.

The Jaina theory of Time may thus be summarised. Time is a real substance, being the accompanying cause of change in other substances. It is either *Vyavahāra Kāla*, empirical time consisting in minutes, hours and other ordinary measures of time, or *Dravya Kāla,* which, although it appears as a heap of jewels, consists really of indefinitely many *Kālāṇus* or time-units. *Kāla* is a *Niṣkriya* substance and has no *Pradēśa* or rather only one *Pradēśa*.

REFERENCES

1. अथ पुद्गल एवात्र धर्मोधर्मो द्विधा नभः।
कालश्च पञ्चधैवेत्यजीव तत्त्वं जगौ जिनः।।

वर्धमान पुराणम् by भट्टारक सकलकीर्ती, १६.१५.

Atha Pudgala Ēvātra Dharmōdharmō Dvidhā Nabhaḥ
Kālaśca Pañcadhaivētyajīva Tattvaṃ Jagau Jinaḥ//

Vardhamāna Purāṇa by Bhaṭṭāraka Sakalakīrtī, 16.15.

धर्मोधर्मो नभः कालः पुद्गलश्चेति पञ्चधा।
अजीव कथ्यते सम्यग् जिनैस्तत्वार्थदर्शिभिः।।

धर्म शर्माभ्युदय काव्य, २१.८१.

Dharmōdharmō Nabhaḥ Kālaḥ Pudgalaścēti Pañcadhā/
Ajīva Kathyatē Samyag Jinaistatvārthadarśibhiḥ//

Dharma Śarmābhyudaya Kāvya, 21.81.

धर्मोधर्मावथाकाशं कालः पुद्गल इत्यपि।
अजीवः पञ्चधा ज्ञेयो जिनागमविशारदैः।।

चन्द्रप्रभ चरितम्, १८.६७.

Dharmōdharmāvathākāśaṃ Kālaḥ Pudgala Ityapi/
Ajīvaḥ Pañcadhā Jñēyō Jināgamaviśāradaiḥ//

Candraprabha Carita, 18.67.

2. वर्तना परिणाम क्रियाः परत्वा परत्वे च कालस्य।

श्री तत्त्वार्थाधिगम सूत्र, V.२२.

Vartanā Pariṇāma Kriyāḥ Paratvā Paratvē Ca Kālasya/

Śrī Tattvārthādhigama Sūtra, V.22.

वत्तणालक्खणो कालो।

उत्तराध्ययन सूत्र, २८.१०.

Vattaṇālakkhaṇō Kālō ।

Uttarādhyāyana Sūtra, 28.10.

3. जीवादीदव्वाणं परिवट्टणकारणं हवे कालो।

नियमसार, ३३.

Jīvādīdavvāṇaṃ Parivaṭṭaṇakāraṇaṃ Havē kālō ।

Niyamasāra, 33.

वत्तणहेदू कालो वत्तनगुण भविय दव्वणिचयेसु।
कालाधारेणेव य वट्टंति हु सव्वदव्वाणि॥

गोम्मटसार, जीव काण्ड, ५६८.

Vattaṇahēdu Kālō Vattanaguṇa Bhaviya Davvaṇicayēsu ।
Kālādhārēṇēva Ya Vaṭṭanti Hu Savvadavvāṇi ॥

Gōmmaṭasāra, Jīva Kāṇḍa, 568.

4. ण य परिणामदि सयं सो ण य परिणामेइ अण्णमण्णेहि।
विविहपरिणामियाणं हवदि हु कालो सयं हेदू॥

गोम्मटसार, जीव काण्ड, ५७०.

Ṇa Ya Pariṇāmadi Sayaṃ Sō Ṇa Ya Pariṇāmēi Aṇṇamaṇṇēhi ।
Vivihapariṇāmiyāṇaṃ Havadi Hu Kālō Sayaṃ Hēd ॥

Gōmmaṭasāra, Jīva Kāṇḍa, 570.

5. दव्व परिवट्टरूवो जो सो कालो हवेइ ववहारो।
परिणामादीलक्खो वट्टणलक्खो य परमट्ठो॥

द्रव्य संग्रह, २१.

Davva Parivaṭṭarūvō Jō Sō Kālō Havēi Vavahārō ।
Pariṇāmādīlakkhō Vaṭṭaṇalakkhō Ya Paramaṭṭhō ॥

Dravya Saṅgraha, 21.

6. नवजीर्णादिपर्यायैर्द्रव्यानां यः प्रवर्तकः।
समयादिमयः कालो व्यवहाराभिधोस्ति सः॥

वर्धमान पुराणाम्, १६.३४.

Navajīrṇādiparyāyairdravyānāṃ Yaḥ Pravartakaḥ ।
Samayādimayaḥ Kālō Vyavahārābhidhōsti Saḥ ॥

Vardhamāna Purāṇa, 16.34.

7. व्यावहारिककालस्य परिणामस्तथा क्रिया।
परत्वं चापरत्वं च लिंगान्याहुर्महर्षयः॥

तत्त्वार्थसार, ३.४५.

Vyāvahārikakālasya Pariṇāmastathā Kriyā।
Paratvaṃ Cāparatvaṃ Ca Liṅgānyāhurmaharṣayaḥ॥

Tattvārthasāra, 3.45.

8. कालो परिणामभवो परिणामो दव्वकालसंभूदो।
दोण्हं एस सहावो कालो खणभंगुरोणियदो॥

पञ्चास्तिकायसार, १०७.

Kālō Pariṇāmabhavō Pariṇāmō Davvakālasambhūdō।
Dōṇhaṃ Ēsa Sahāvō Kālō Khaṇabhaṅgurōṇiyadō॥

Pañcāstikāyasāra, 107.

9. क्रिया दिनकरादीनामुदयास्तमयादिकम्।
प्रविहायापरः कालो नास्तीत्येके प्रचक्षते॥

चन्द्रप्रभ चरितम्, १८.७५.

Kriyā Dinakarādīnāmudayāstamayādikam।
Pravihāyāparaḥ Kālō Nāstītyēkē Pracakṣatē॥

Candraprabha Carita, 18.75.

10. लोकाकाशप्रदेशे ये ह्येकैक अणवः स्थिताः।
भिन्नभिन्नप्रदेशस्या रत्नानामिव राशयः॥

वर्धमान पुराणाम्, १६.३५.

Lōkākāśapradēśē Yē Hyēkaika Aṇavaḥ Sthitāḥ।
Bhinnabhinnapradēśasyā Ratnānāmiva Rāśayaḥ॥

Vardhamāna Purāṇa, 16.35.

11. दोण्हमणूमदिक्कमकालपमाणं हवे सो हु॥

गोम्मटसार, जीव काण्ड, ५७३.

Dōṇhamaṇūmadikkamakālapamāṇaṃ Havē Sō Hu॥

Gōmmaṭasāra, Jīva Kāṇḍa, 573.

CHAPTER - III

THE CONCEPT OF TIME IN BUDDHISM

In the history of Indian religions, *Buddhism* occupies a unique place. The teachings of Lord *Buddha* form the main crux of Buddhism i.e., the four noble truths *(Duḥkha, Duḥkha Samudāya, Duḥkha Nirōdha, Duḥkha Nirōdha Mārga)* leading to *Aṣṭāṅga Mārga* based on Right Views, Right Resolve, Right Speech, Right Conduct, Right Livelihood, Right Effort and Right Concentration. So, all these will hold the individual to attain the higher stage of life called the *Nirvāṇa.*

During the early period of Indian thought, Time *(Kāla)* baffled the Indian thinkers to such an extent that they came to look upon it not only as the cause of the Universe but also as an all-pervading principle which governs everthing in it.[1] But as speculation advanced, Time came to be considered as one of the causes which determines the course of natural phenomena. Thus, a later scholiast *Śilāṅka,* attempting to explain the non-absolutism in Jaina thought, maintained that Time can be recognized as one of the causal factors in the evolution of nature, because it is found that certain events like the flowering of trees, etc., occur at certain times only, not all the time.[2] In the thought of the *Upaniṣads,* with its emphasis on permanence and eternity and the resulting denial of change and causation as being illusory, the problem of causation received scant attention. But in early Buddhism where there is a denial of permanence, the conceptions of change and causality of time occupied prominent places.

Causality *(Pāli, Praticcasamuppāda; Sanskrit, Pratītya-samutpāda)* is the central philosophy of early Buddhism. With the insight he gained as he sat contemplating under the Bōdhi tree on the bank of the river *Nerañjara,* the Buddha realized that everything in this world of experience is causally produced. It was this insight, that enabled him to eliminate all the doubts *(Kaṅkha)* he had entertained regarding the nature of existence.[3] Thus, in early Buddhism, the recognition of the fact that everything is impermanent *(anicca),* conditioned *(Saṅkaṭa),* and causally produced *(Praticcasamupanna),* along with the denial of anything permanent *(nicca)* or eternal *(dhuva, sassata)* led to fruitful speculation regarding Time. But it must be emphasized that during this early period of Buddhist thought whatever reflections there were on the problem of time were invariably connected with speculation on the nature of things which are temporal *(Kālika).*

The Buddha's conception of the Universe, unlike his later *Mahāyāna* disciples who maintained that there is no beginning *(Anavarāgra)* of the world process *(Saṃsāra),*[4] is that the beginning is totally inconceivable *(Anamatagga).*[5] Although the beginning is inconceivable, yet, it is possible to see periods of evolution *(Vivaṭṭa)* and dissolution *(Saṃvaṭṭa)* following one another.[6] These periods are reckoned in terms of aeons *(kappa),* each of which is said to be of such immense duration that it can only be illustrated by means of similes, and these became very popular in Indian thought later on. One such simile is as follows; "If there were to be a great mountain, one league in width, one league in length and one league in height, a solid mass without chasms or clefts, and a man at the end of every hundred years were to strike it once each time with a silken cloth, that mountain will sooner be done away with than would an aeon".[7] This passage, while emphasizing the immensity of time, also shows that we can observe events and use processes based on these events to measure time.[8] Thus, we derive Time from events which are in Time *(Kālika)*, but not vice versa.[9]

The processes of evolution and dissolution are said to take place in accordance with the causal principle which is stated as : "When this exists, that exists or comes to be; on the arising of this, that arises. When this doesn't exist, that doesn't exist or come to be; on the occasion of this, that ceases".[10] Since it has been pointed out that, "there is a profound connection between the reality of time and the existence of an incalculable element in the universe".[11] The causal principle has several theories, one of which was popular in the early *Upaniṣadic* tradition. In the early Buddhistic texts, this is referred to as the theory of self-causation *(Sayaṅkata-Vāda)*[12] and was based on the belief in a self *(Attan)*, considered to be the essence of everything *(Sabbam)*. Thus, "everything exists" *(Pāli, sabbam atthi;* Sanskrit, *Sarvam Asti)*[13] means that this essence of everthing exists. This leads to the view that the consequence pre-exists in the cause, the future in the present. Such a strictly determined causal principle would also mean that we can by examining the present, predict with absolute certainty what will happen in the future, for the future is merely the hidden present. For this reason temporality becomes a mere illusion.

The Buddha's theory of causality differs from this in that it is not a form of strict determinism of this sort. The very use of the term "when" (and this idea is expressed by the use of the locative absolute construction in *Pāli*), which in this context is equivalent to the conditional particle 'if', should be sufficient to show that the future events cannot be predicted with absolute Certainty. For this reason, we find that none of the extrasensory perceptions recognized in early Buddhism refer to the future. As is well known, Omniscience *(Sabbaññuta)*, as later understood, was not claimed by the Buddha. We come across only two instances when the Buddha made any kind of prediction into the future with much certainty. One is the prediction that a 'Stream entrant' *(Sotāpanna)* is certain *(Niyata)* to attain enlightenment *(Sambōdhi Parāyana)*,[14] and the other is that a person who has eliminated craving and thus attained enlightenment will not be reborn *(Khīṇa Jāti, Nāparaṃ Itthattāya)*.[15] Both these would be considered knowledge based on inductive inference *(Anavaye Jñana)*.[16] Depending on such inductive inferences the Buddha recognized the future validity of the causal principle. This prompted him to make such declarations : "Whether the *Tathāgatas* were to arise in this world or were not to arise, this causal status, this causal pattern, this conditionality remains".[17] If this causal pattern is said to exist always in this world, and if "perceiving the *dhamma* means perceiving causality *(Patīccasamuppāda)*,"[18] then it is possible to say that his teachings will be valid at all times. It is only in this sense that the teaching *(dhamma)* was described as being timeless *(Akālika)*.[19] On the basis of this theory causality, it is possible to define the three periods of time, past, present and future in the following manner; the past is the determined *(Bhūta)*; the present is the moment of becoming *(Bhava)*; and the future is the as yet determined *(Bhavya)*.[20]

Thus using the terminology of Whitrow on the problem of Time, it is possible to say that according to Buddhism, "the future is hidden from us—not in the present, but in the future",[21] Time is the mediator between the past and the present. Hence for that which is born *(Jātassa)*, death is a matter of time. This may be the idea behind the use of the phrases *Kālaṃ Karōti*[22] literally, 'he does his time', 'Is fulfilling his time', which means dying, and *Kālaṃ Kaṅkhati*[23] literally, 'awaiting time', that is awaiting death. For this reason, Time assumes the position of *Māra,* the personification of Death. The famous quatrain in the *Jātaka* runs :

Time consumes all beings
including oneself;
The being who consumes time,
cooks the cooker of beings.[24]

Although time is supposed to overwhelm ordinary human beings, yet the one who has attained enlightenment is able to bring time under his control. Just as he overwhelms *Māra,* the evil one, even so does he overcome time. He is said to overcome time, not because he attains a state of permanent existence (as it was advocated in *Mahāyāna),*[25] but because of the complete eradication of craving and attachment he no longer has any longing for existence or anything associated with it. Hence, dying or 'fulfilling time' *(Kālakiriyā)* never worries him, as it continued becoming *(Bhava)*. Thus, immortality *(Pāli, Amata;* Skt. *Amṛta)* in early *Buddhism* becomes a synonym of no-rebirth *(A-punabbhava)*.[26] He who has overcome the process of becoming also overcomes time, because there is no time apart from the process of becoming.

It was mentioned that during the pre-Buddhist period there was a theory which upheld that "everything exists". The foundation of this theory seems to be the view that an event can never cease to be an event.[27] Because of the popularity this theory enjoyed during this period, the Buddha seems to have taken much trouble to refute it. In addition to the argument from experience that he adduced to refute this theory, the Buddha also pointed out that it was the result of a linguistic fallacy.[28] His criticism runs thus : Monks, there are these three linguistic conventions or usage of words or terms which are distinct, have been distinct in the past, are distinct at present and will be distinct in the future and are not ignored by the recluses and Brahmans who are wise. Which three? Whatever form *(Rūpa)* there has been, which has been ceased to be, which is past and has changed is called, reckoned or termed "has been" *(Ahosi);* it is not reckoned as 'it exists' *(Atthi)* nor as 'it will be' *(Bhavissati)*. (The same is said about the other four aggregates—Sensation, Perception, Dispositions and Consciousness) .....Whatever form is not arisen, not come to be, is called, reckoned or termed "it will be" *(Bhavissati)* and is not reckoned as "it exists" or as "it has been". .....Whatever form has become and has manifested itself is called, reckoned or termed as "it exists" *(Atthi)* and is not reckoned as it has been or as "it will be".[29]

It is very clear from this passage that the theory according to which the past and the future exist in the present or even the view that a thing exists during the past, the present and the future are the results of unwarranted interpretation of linguistic usage. Here the temporal Copulas "was", "is now" or "will be" in prepositions which assert temporal relations between events are being confused with the timeless Copulas of logic.

One of the arguments against the reality of time seems to be based on the view that it is not an object of the five types of sense experience. Hence it is nothing but a mysterious illusion of the intellect.[30] For early Buddhism, which not only recognized six senses but also held that the sixth sense (Mind) could be developed to perceive, with the help of memory, a long period of one's past history time was an important ingredient or experience.[31] As if rejecting the theory of atomic moments of time, which is recognized to be abstraction,[32] the Buddha is represented as saying : "This physical body made up of the four primary existents is seen to exist for one, two, three, four, five, ten, twenty, thirty, forty, fifty, hundred or more years. That which is called the mind, thought or consciousness arises as one thing and ceases as another whether by night or by day".[33] This is a clear recognition of the two types of experience that we have. One is the experience of things which endure for some finite segment of time, and the other is the experience of momentary things. The latter experience is illustrated by the simile of the flowing river. According to the description in the early Buddhist Texts, 'there is no moment, no linking, no particle of time that the river stops flowing'.[34] These two types of experience of time can be compared to the experienced one has of the movement of the two hands of the clock, the hour hand, the second hand.[35] It is important to note that early Buddhism never reduced the experience of the former to the experience of the latter, although logical abstraction can always lead to such a hypothesis. In fact, a reduction of every form of experience into further indivisible moments *(Pāli, Khaṇa; Skt. Kṣaṇa)* was carried out during the period of scholasticism, and such a theory is conspicuous by its absence in the early discourses. Therefore, during the early period, although there were statements which could be interpreted to mean a theory of momentariness, the most dominant view was the one which recognized a finite segment of time constituting our immediate experience. The recognition of this finite segment according to early Buddhism there is a duration of temporal experience with a certain unification of perspective. For this reason, unlike in the cases where a theory of moments dominated,[36] early Buddhism considered both time and causation as parts of our experience, not as mere inferences bases primarily on the succession of momentary ideas. Thus, it is possible to maintain that early *Buddhism* presents us with an empiricist analysis of time.

With regard to the problem of time, early *Buddhism* seems to have followed the middle path, so famous in the history of Buddhist thought. It appears as if it considered absolute time as an extreme and an unnecessary hypothesis. The other extreme is the consideration of time as a mysterious illusion of the intellect. Avoiding both these extremes, the *Buddha* seems to have considered Time as an essential feature of the Universe and the experience of it.

With the development of Scholasticism after the passing away of the *Buddha,* this empiricist philosophy of Time and temporality changed completely. Unbridled speculation during the period of the *Abhidharma* led to the development of many theories which are more metaphysical than empirical.

In his desire to eliminate the Pre-Buddhist conception of an eternal and immutable self *(Attan),* the *Buddha* adopted the analytical method *(Vibhajjavāda)* of reducing things to their components. Thus, the human personality was analyzed sometimes into five aggregates *(Pāli, Khandha; Skt. Skandha)* and at other times into six elements, *(Dhātu)*. Yet, all these aggregates as well as the elements, were the contents of experience, not of pure logical analysis. But with the development of scholasticism, this analytical approach was carried to its logical conclusion, and the result was the emergence not only of a theory of atoms *(Paramāṇu)* but also a theory of moments *(Khaṇa)* spatial analysis giving rise to temporal atomicity.[37] The development of these two theories occasioned several other doctrines which are not compatible with the basic teachings of the Buddha.

One of the immediate results of the analysis of time into atomic units or chronons was the view that time is fleeting or flowing from the future into the present and from the present into the past. Thus, in the commentarial literature, we come across expressions such as *Gacchante Gacchante Kāle* meaning "as time passed by" or "with the passage of time."[38]

All these developments were summarized by *Buddhaghōṣa* in the *Atthasālini,* his commentary on the *Dhammasaṅganī,* where he refers to the three varieties of the present *(Paccupanna).*[39] The first is the protensive present *(Aādhā Paccuppanna)* or the finite segment of time constituting our immediate experience. This is said to be the kind of present recognized in the discourses *(Sutta).* The second is the flowing or the continuing present *(Santati paccuppanna),* which was then accepted in the commentarial tradition *(Aṭṭhakatha).* The third is the momentary present *(Khaṇa paccuppanna),* which is not identified with any tradition. Yet, it was the conception mentioned last which dominated the Scholastic tradition.

Buddhaghōṣa's commentary on the *Vibhaṅga,* the second book of the *Abhidhamma Pīṭaka,* in introducing the theory of moments, says that it is a doctrine peculiar to the *Abhidhamma* and not to the *Suttanta.* According to this, *Rūpa* (matter or form) is classified as past, future, or present *(Atītānāgata paccuppanna)* in the *Suttanta* and this division is made on the basis of becoming *(Bhavēna paricchinna).* In the *Abhidhamma,* however, the division is made on the basis of moments *(Khaṇēṇa Paricchinna).*[40]

There are different versions of the theory of moments. The *Sarvāstivādins* recognized four moments, the nascent *(Jāti),* the static *(Sthiti),* the decaying *(Jara),* and the cessant *(Nāśa).*[41] Post-*Buddhaghōṣan Thēravāda* recognized three : the nascent *(Uppāda),* the static *(Ṭhiti),* and the cessant *(Bhanga).*[42] The *Sautrāntikas* differed from all of them in accepting two moments only, the nascent *(Utpāda)* and cessant *(Vyaya)*, and rejecting the static moment *(Sthiti-Kṣaṇa)*.[43]

The difficulties posed by the analysis of time into atomic units are numerable. Although a moment was considered to be the unanalizable unit of time, it was found that a distinction had to be made between a moment of thought and a moment of matter, because, as pointed out earlier, thought changes more rapidly than physical bodies.[44] Therefore, in the *Thēravāda Abhidhamma* it is said that during the life time of a single moment of matter sixteen moments of thought arise and pass away.[45]

The most difficult problem created by theory of moments concerned the experienced continuity of temporal events. A moment was considered to be durationless, comparable to the dimensionless point of space, and hence, past, present and future moments are utterly distinct from each other. They are discrete. To explain the problem of continuity arising as a result of the analysis of time into momentary and discrete units, the Buddhist schools offered various solutions. The most widely known and the most severely criticized of these different solutions was that proffered by the *Sarvāstivādins* who insisted that underlying the succession of momentary events is the substance or "own nature" *(Dravya, Svabhāva)* which remains unchanged.[46] Thus, a thing of Dharma has two aspects; the characteristic *(Lakṣaṇa)* which is temporal *(Kālika, Kṣaṇika)* and the substance *(Dravya)* which is eternal or timeless. The term *"Sarvāstivāda"* means the 'theory that everything exists" *(Sarvam asti)*. This everything (Sarvam) was the substance of everything which, they held, exists during the three periods, past, present and future. This theory was examined by the *Thēravādins* in their *Kathāvatthu* where they make the *Sarvāstivādins* admit that not only everything past, present and future exists, but that past, present and future themselves exist, that is, they are independently real.[47] While the basic theory of the *Sarvāstivādins* was not much different from *Upaniṣadic* theory of "everything exists" referred to and criticized by the *Buddha,* there is also a significant difference, unlike the thought of the *Upaniṣads* which is idealistic, the *Sarvāstivāda* represented as schools of realism and therefore, as pointed out by the *Thēravādins,* they upheld the independent reality not only of things, but also of Time. It is interesting to note that in support of their rejection of this theory of "everything exists" *(Sarvam asti),* the *Thēravādins* are represented as quoting a discourse from the Buddha on the nature of linguistic conventions referred to earlier.[48]

The school of Buddhism known as the *Sautrāntika* rejected the *Sarvāstivāda* conception of substance *(Svabhāva)* as being no different from the theory of self *(Ātman).*[49] But the rejection of this underlying substance compelled them to the view that there is no duration whatsoever and that what appears as duration is a series of fleeting moments, like the Cinematograph. This continuum of durationless instants, no doubt, is the result of logical abstraction, a theoretical construction based on the empirical data of consciousness. Unlike the *Thēravādins* who recognized two different types of moments, that is, a moment of thought and moment of matter, the *Sautrāntikas* made no such distinctions. The most important consequence of this conception of time was the theory that there are no instantaneous connections between external events and the observer.[50] Hence, there is no direct perception of an object; there is inferential knowledge *(Anumāna)* only. Thus, the *Sautrāntikas* were popularly known for their doctrines of "Representationism", that is, the inferability of the external object *(Bāhyārthānumēya Vāda)*.[51]

In the *Mūla Madhyamakārika, Nāgārjuna* devoted one chapter to the examination of the problem of time.[52] It is indeed a very short but extremely important chapter. There is no doubt that it was the *Sarvāstivāda* conception of Time which drew the criticism from *Nāgārjuna.* The theory, as explained by *Candrakīrti* recognized by the existence of the substance of things during the three periods. It was argued by the realistic that since substance exists, time with which it is associated also exists.[53] It is this independent reality of time which *Nāgārjuna* takes up for criticism. His criticism was based on the fact that a thing *(Bhāva)* or its substance *(Bhāvasvabhāva)* and time *(Kāla)* are relative to or dependent upon one another. Early *Buddhism* as pointed out earlier, would have stopped at this point, but *Nāgārjuna* did not. He employed his dialectic to demonstrate the unreality of non-existence of Time.[54] *Nāgārjuna's* argument was based on the assumption that two things cannot be related unless they are co-existent. Hence, if present and future are held to be contingently related to the past, then both present and future should be in the past. Otherwise they cannot be contingently related. On the other hand, present and future could not exist without being contingent on the past. Hence, according to *Nāgārjuna,* there is no justification for the recognition of a present and a future time.[55] The same agreement was used to refute the reality of the past. Past, present and future which were comparable to such concepts as above, below and middle *(Uttamādhamamadhyama),* or arising, enduring, and passing away *(Utpādasthitibhaṅga)* etc., could not thus withstand the onslaught of *Nāgārjuna's* dialectic.

The opponents argued that Time exists because it has

dimension and is measured in terms of moments, days, fortnights, months and years.[56] Nāgārjuna pointed out that it is not possible to measure Time. He maintained that non-enduring or non-static Time cannot be measured, because it cannot be manipulated, and that an enduring or static time although manipulatable, does not exist.[57] By way of conclusion, *Nāgārjuna* points out that if time exists depending on existential structure is to be found, for he has already refuted such a structure.[58] Hence, according to him, Time does not exist.

Nāgārjuna claimed that his exposition of the doctrine is based on the recognition of the two truths, the conventional *(Saṃvṛti-Satya)* and the ultimate truth *(Paramārtha-Satya)*.[59] According to *Candrakīrti,* the ultimate truth is the independent *(Aparapratyaya),* peaceful *(Śānta),* without conceptual proliferation *(Prapañcātīta),* non-conceptual *(Nirvikalpa),* and without plurality *(Anēkārtha)*.[60] Hence, *Nāgārjuna's,* no doubt, was a transcendentalist criticism of phenomenal reality. This would become clear from a comparison of the standpoints adopted by the two schools, early *Buddhism* and *Mādhyamika.* When early *Buddhism* maintained that things are relatively real, *Nāgārjuna* insisted that things are unreal because they are relative, the real being the transcendental.

The Buddhistic Concept of Time may thus be summarized : 1) early Buddhism presented an empiricist and relativistic conception of Time; 2) the *Abhidhamma* Scholasticism produced an absolutistic conception mainly because of its speculative approach and 3) the *Mādhyamikas,* as a result of their transcendentalism, denied the reality of Time.

REFERENCES

1. कालादापः समभवन् कालाद् ब्रह्म तपो दिशः।
कालेनोदेति सूर्यः काले नि विशते पुनः॥
अथर्ववेद, १९.५४.१.
Kālādāpaḥ Samabhavan Kālād Brahma Tapō Diśaḥ।
Kālēnōdēti Sūryaḥ Kālē Ni Viśatē Punaḥ॥
Atharvavēda, XIX.54.1.
कालेन वातः पवते कालेन पृथिवी मही।
द्यौर्मही काल आहिता॥२॥
Kālēna Vātaḥ Pavatē Kālēna Pṛthivī Mahī।
Dyaurmahī Kāla Āhitā॥2॥
कालो ह भूतं भव्यं च पुत्रो अजनयत् पुरा।
कालादृचः समभवन् यजुः कालादजायत॥३॥
Kālō Ha Bhūtaṃ Bhavyaṃ Ca Putrō Ajanayat Purā।
Kālādṛcaḥ Samabhavan Yajuḥ Kālādajāyata॥3॥
कालो यज्ञं समैरयद् देवेभ्यो भागमक्षितम्।
काले गन्धर्वाप्सरसः काले लोकाः प्रतिष्ठिता॥४॥
Kālō Yajñaṃ Samairayad Dēvēbhyō Bhāgamakṣitam।
Kālē Gandharvāpsarasaḥ Kālē Lōkāḥ Pratiṣṭhitāḥ॥4॥
कालेयमङ्गिरा देवोथर्वा चाधि तिष्ठतः।
इमं च लोकं परमं च लोकं पुण्यांश्च लोकान् विधृतीश्च पुण्याः।
सर्वाल्लोकानभिजित्य ब्रह्मणा कालः स ईयते परमो नु देवः॥५॥
Kālēyamaṅgirā Dēvōtharvā Cādhi Tiṣṭhataḥ।
Imaṃ Ca Lōkaṃ Paramaṃ Ca Lōkaṃ Puṇyāṃśca Lōkān Vidhṛtīśca Puṇyāḥ।
Sarvāllōkānabhijitya Brahmaṇā Kālaḥ Sa Īyatē Paramō Nu Dēvaḥ॥5॥
Atharva Vēda Saṃhita, ed. R. Roth and W.D. Whitney (Berlin : F. Dummler, 1924), 19.54; Maitri Upaniṣad, 6.14.
2. Sūtrakṛtāṅga with the commentary of Śilāṅka, ed. A.S. Suri and C. Gaṇīndra (Bhāvanagara, Bombay, Vijayadēva Sūra, 1950-1953), I.31-2, *"Tathā Kālō Pi Karta yatō Bakulacampakāśōka Punnāga Nāga Sahakārādīnāṃ Viśiṣta Ēva Kālē Puṣpaphalādyubhavō Na Sarvadēti"*.
3. Udāna, ed. P. Steinthal (London : PTS, 1948, reprint), p. 1 :

"Yadā Havē Patubhavanti Dhammā | Ātāpinō Jhāyatō Brahmaṇassa |
Athassa Kaṇhā Vāpayanti Sabbā | Yatō Pājānāti Sahētudhammaṃ" |

4. Mūlamadhyamakakārikas, des Nāgārjuna, avec la Prasannapada commentaire de Candrakīrti, publie par Louis de la Vallee Poussin (St. Petersburg : Academy of Sciences of USSR, 1903-1913), XI.1.
5. Samyutta Nikāya, ed. Leon Feer (London : PTS 1884-1904) I, 178 ff.; Sāratthappakāsini, ed. F.L. Woodward (London : PTS, 1929-1937), II, 156, Anamattaggo aviditaggo.
6. Dīgha Nikāya ed. T.W. Rhys Davids and J.E. Carpenter (London : PTS, 1890-1911), III, 84-5; Taiṣō Ṣinṣu Daisōkyō (herafter cited as Taisho), ed. J. Takakusu and K.Watanabe (Tokyo, The Taisho Shinshu Daizokyo Kanko Kai, 1962, reprint), I, 37 b-c.
7. Samyutta Nikāya, II, 181-4; Taisho, II, 242a-243b.
8. Whitrow, G.J., The Natural Philosophy of Time (London : Thomas Nelson & Sons. 1961), p. 34.
9. Cp. J. Alexander Gunn, the Problem of Time (London, 1929), p. 323 (quoted by Whitrow, op. cit., p. 36).
10. Udāna, p. 1; Majjhima Nikāya, ed. V. Trenckner and R. Chalmers (London : PTS, 1948-1951), I, 262-4; Samyutta Nikāya, II, 28, 70, 96; Candrakīrti, 9; Taisho, I, 562c; II, 10a, 67a, 713c-714a.
11. Whitrow, op. cit., p. 295.
12. Samyutta Nikāya, II, 18; Taisho, I, 76 a; II, 81 a, 86 a-c.
13. Samyutta Nikāya, II, 17; Taisho, II, 85c.
14. Dīgha Nikāya, I, 156; Taisho, I, 616c.
15. Majjhima Nikāya, I, 184; Taisho, I, 658a.
16. Samyutta Nikāya, II, 58; cp. Taisho, II, 99c-100a.
17. Samyutta Nikāya, II, 25; Taisho II, 84b; Candrakīrti, 40.
18. Majjhima Nikāya, I, 190-1; Taisho I, 467a, "Yō Paṭiccasamuppādaṃ Passati Sō Dhammam Pasati."
19. Aṅguttara Nikāya, ed. R. Morris and E. Hardy (London : PTS, 1885-1900), III, 212, etc.; Taisho, I, 616c.
20. Cp., op. cit., p. 295.
21. The Natural Philosophy of Time, p. 259.
22. Vinaya Pīṭaka, ed. H. Oldenberg (London : PTS, 1879-1883), III.80.
23. Samyutta Nikāya, I, 187; Sutta Nipāta, ed. D. Anderson and H. Smith (London; PTS, 1948), p. 516.
24. Jātaka, ed. V. Fausboll (London, 1895-1907), II, p. 260.
25. कालो घसति भूतानि सब्बान् एव सहत्तना
    यो च कालघसो भूतो सो भूतपचनिं पचि। Cp. मैत्रायणी उपनिषद्, ६.१५.
    Kālō Ghasati Bhūtāni Sabbān Ēva Sahattanā

Yō Ca Kālaghasō Bhūtō Sō Bhūtapacanim Paci I

Cp. Maitrāyaṇī Upaniṣad, 6.15.

26. Saṃyutta Nikāya, I, 174; Taisho, II, p. 27b.
27. See J. M. E. Mctaggart, Nature of Existence (Cambridge : At the University Press, 1927, II, p. 10.
28. Samyutta Nikāya, II, 17; Taisho, II, p. 85c.
29. Cp. C.D. Broad, Examination of Mctaggart's Philosophy (Cambridge : At the University Press, 1938), II, p. 316.
30. Whitrow, op. cit., p. 313.
31. Dīgha Nikāya, III, 134; cp. Taisho, I, p. 75 b-c.
32. Whitrow, op. cit., p. 312.
33. Samyutta Nikāya, II, 94-7; Taisho, II, 81c-82a.
34. Aṅguttara Nikāya, IV, 137; cp. Taisho, I, p. 682b.
35. C.D. Broad, Scientific Thought (London : Routledge & Kegan Paul, 1923), p. 351.
36. A Treatise of Human Nature, ed. L.A. Selby-Bigge (Oxford : The Claredon Press, 1888), p. 39f.
37. Whitrow, op. cit., p. 153.
38. Dhammapada-Aṭṭhakathā (London : PTS 1906-1914), I, 319.
39. Atthasālini, ed. E. Muller (London : PTS, 1987), p. 421.
40. Sammōhanavinōdini, ed. A.P. Buddhadatta (London : PTS, 1923), p. 7.
41. Abhidharmadīpa, ed. P.S. Jaini (Patna, K.P. Jayaswal Research Institute, 1959), p. 104.
42. Sammōhanavinōdini, 7.
43. Sphuṭārthābhidharmakōśavyākhya of Yaśōmitra, ed. U. Wogihara (Tokyo : The Publication Association of Abhidharmakōśavyākhya, 1932-1936), p. 33.
44. Sammōhanavinōdini, 25, Rūpaṃ Garupariṇāmaṃ Dandhanirōdhaṃ, Arūpaṃ Lahupariṇāmaṃ Khippanirōdhaṃ.
45. *Ibid.*, 25-26, Rūpē Dharēnte Yēvá Sōḷasa Cittāni Uppajjitvā Nirujjhanti, Taṃ Pana Sattarasamēna Cittēna Saddhiṃ Nirujjhati.
46. L'Abhidharmakōśa Dē Vasubandhu, traduction et annotations par Louis de la Vallee Poussin (Paris : Paul Geuthner, 1923-1925), V.52 ff. Taisho, XXIX, 104c; Ad. 259; Tattvasamgraha, ed. E. Krishnamacharya (Baroda : Oriental Institute, 1926), I, 504.
47. Points of Controversy (translation of Kathāvatthu), by S.Z. Aung and Mrs. Rhys Davids (London : PTS, 1915), p. 84ff.
48. *Ibid.*, p. 95f.
49. Sphuṭārthābhidharmakōśavyākhya of Yaśōmitra, p. 362.

50. Whitrow, op. cit., p. 179, where it is given as a theory formulated by Albert Einstein.
51. Sarvadarśanasaṅgraha, ed. V.S. Abhyankar (Poona : Bhandarkar Oriental Research Institute, 1924), p. 36.
52. Kālaparīkṣa, Chap. XIX.
53. Candrakīrti, 382.
54. Mūlamadhyamakakārikas, XIX, 3, Tasmāt Kālō Na Vidyatē.
55. *Ibid.*, XIX, 1.
56. Candrakīrti, 385.
57. Mūlamadhyamakakārikas, XIX, 5.
58. See Mūlamadhyamakakārikas, Chapters IV, V, VII, XV and XVIII.
59. *Ibid.* XXIV, 8.
60. Candrakīrti, p. 493.

CHAPTER - IV

COSMOLOGICAL AND HISTORICAL CONCEPTS OF TIME

It is a widely, almost unusually held view that notion of Time known to early India is cyclical as opposed to the notion of Time in the Semitic religions-Judaism, Christianity and Islam wherein it is linear. But historical consciousness did exist in early India. Both cyclic and linear time were used, but their functions differed. Cyclic time occurs frequently in cosmological context such as Hindu theory of *Yugas, Manvantaras,* and *Kalpas*. Whereas linear time in historical sources such as Genealogies, Biographies and Chronicles, where time reckoning was recorded through generations, regnal years and eras.

Both cyclic and linear times are at one level linked with ideas of Cosmology—the theory of Universe as an ordered whole and of the possible laws, which govern it. Therefore, Cosmological time relates to the Universe is almost infinite and where it is cyclic consisting constructions of many cycles of time setting out as it were a calendar for the Universe.

Initially those who make the Calendar control the reckoning of time. Calendars are part of a larger system of time measurement and calendar makers were proficient in Astronomy. Concept of time and theories of Astronomers sometimes run parallel, although they are necessarily inter-dependent. Astronomers calculated time and conceptualized it. This took the form of extensive computations relating to planetary distances, orbit, asterisms, eclipses and such like, the accumulated knowledge of endless years of observing the night skies and daily shadow reading. These could become patterns in cosmic time. Cosmological time need not be either complete fantasy, or intuitive. It could draw on the existing notions of time current in the calculations of Astronomers and Mathematicians, although imagination would also be brought into play in the final construct.

Units of time are often initially natural units such as the cycle of day and night, the cycle of the month based on phases of the Moon—this synodic and lunar fortnights and the seasonal cycle of the year, starting generally with the Spring equinox with segments named after the seasons : What are referred to in the early Indian texts as *Ahōrātra, Māsa* and *Pakṣa* and the *Ṛtus*. The year was also divided into the Northern and Southern course of the Sun on the basis of the solstice—*Uttarāyaṇa* and *Dakṣiṇāyana*. The Lunar day or *Tithi* consisted of *Muhūrtas* sometimes refined further to finer measurements as minute as the blinking of an eye.

What may be called ritual time, revolved around the cycles of nature. Seasonal rituals often evolved from the routine activities of a given society over the year, e.g., the grazing circuits of herders or the sowing and harvesting periods of cultivators. Some of these were also moments for the gathering of wealth. Ritual time is broadly predictable. Ritual texts some times describe the year in terms of the *Seasons*.[1] It comes to be treated as more precise when it is tied to Astrology. With reckoning in *Muhūrtas,* for example, the notion of the auspicious moment comes into existence. Where the ritual is meticulously observed, it suspends the performers of the rituals in a threshold condition where only the parameters of their reckoning prevail. Cyclical theories of time arise from the observance of rhythm based on the sequences in relation to the Sun and the Moon or the *Seasons.*[2] The concept of *Ṛta* as a laggering the Universe to ensure regularity and predictability also derives from the notion of rhythm.

In the early Indian literature where the need to determine time was also linked to the efficacy of sacrificial rituals. Time is described as a five spoked wheel, constantly revolving, evoking regular spacing and cyclic movement.[3] In a five year cycle the Solar and Lunar calendars were adjusted.

The *Nakṣatras* or constellations called for particular attention and this remained a significant activity in early Indian Astronomy. The path of the Sun and the Moon were marked on the basis of a stellar frame. Initially, the five year cycle was referred to as a *Yuga* although later it was extended to a much longer period. The term *Yuga,* literally a yoke is intended to suggest a binding together as an entity. Its usage connected with time refers to a period in which planetary bodies are in conjunction both at the start and at the end. In the early period the major planets were of course the *Sun* and the *Moon* which had already figured in Stellar Astronomy. A far more extensive span of time than a five-year cycle was required with planetary movements becoming important. The *Yuga* carried not only the notion of a natural cycle and was therefore benign and harmonious but the conjunction of the planets carried another meaning in that it suggested a variety of bipolarities-good and evil, divine and human, life and death.

The reckoning was generally based on a Luni-solar calendar. The earliest sense of a calendar had to do with time makers, both individual life cycle and involving the environment, which were gradually ritualized. Some had been established in accordance with the Lunar calendar with easily comprehensible calculations based on the phases of the Moon. The precision of the Solar calendar was useful in agricultural activities and also in Horoscopy and in either case it became an agency of social control. Measurements of time were required by Astrology in making of horoscopes. The urgency to determine time now included the need to know the proper moment for ceremonies and observances-*Saṃskāras*—Particularly important to the identity of upper castes—as indeed also to improve upon calendars and horoscopes. This became an aspect of ritual time, which soon evolved its own forms and rules.

Indian interest in Astronomy is revealed in texts dating to the 5th century B.C. and often included in the category of *Jyōtiṣa Śāstra*. The measurement of time was an important aspect in these studies. The basic unit was initially the five-year *Yuga*. Infusions from West Asia introduced further calculations of cyclic time has developed in Greek Astronomy based in turn on Babylonian ideas. This interest was an indirect result of the substantial trade and contacts both overland and maritime between the Eastern Mediterranean and Hellenistic West Asia with Northern and Western India. *Yavanas* from two former areas were the visiting traders and some seem to have settled for long periods in India. The mingling of Indian and Greeco—Babylonian ideas enhanced activity in Astronomy and Mathematics in India. This took the form of extensive calculations relating to planets, orbits, eclipses and the like. It is said that there was a radical change after 400 A.D., when the *Siddhānta Jyōtiṣa* replaced the *Vēdāṅga Jyōtiṣa* and calculations earlier based on Stellar and Lunar observations now preferred to incorporate planetary motions and Solar reckonings.

Cosmological Time

In the construction of the large cycles of Cosmological time the figures used both in Cosmology and in Astronomy come to be central. In what has been called the *Yuga* Astronomy of the fifth century A.D., Indian astronomers calculated that a *Kalpa,* the largest period of time, consisted of 4320 million years. The astronomers may have borrowed the notion of a *Kalpa* from Purāṇic sources since they required a long period of time as the basis of their calculations. This raised two questions, one relating to the current theories of Cosmological time and the other concerning the dialogue between the authors of a variety of texts.

Cosmological time is described in *Manu's Dharmaśāstra* and in the *Mahābhārata* and is further elaborated upon in the *Purāṇas.* Concepts of time, integrated with ideas on creation, go into the making of what might be called Cosmological time. *Manu* quotes the *Maharṣi Bhṛgu* in describing these concepts which are linked to the notion of the *Yugas* calculated in terms of humanly manageable time, as well as the larger unit of Divine time associated with *Brahma.*[4] It is said that there were six *Manus* before the present one and this provides one kind of time scale. This is followed by a detailed working out of the smaller units of time from the blinking of an eye to a human year with their equivalents in the enhanced time scale used by the *Pitṛs/ Ancestors* and the *Dēvas/Gods.* Given the propensity to classification, different time units were associated with a hierarchy of persons in ascending orders of magnitude. A human month is a day and a night for the ancestors and a human year is the same for the Gods. Manu continues with the description of the four *Yugas* : the first, the *Kṛta* lasts for four thousand years with a preceding and subsequent twilight period of four hundred years each. The next three, the *Trēta, Dvāpara,* and *Kali,* are calculated by reducing one thousand from each with a corresponding reduction of one hundred in the twilight period.

Kṛta	4000+(400×2)	= 4800
Trēta	3000+(300×2)	= 3600
Dvāpara	2000+(200×2)	= 2400
Kali	1000+(100×2)	= 1200
Total	10000+(1000×2)	= 12000

The total of twelve thousand years constitutes an age of the Gods. A thousand of these constitute a single day of *Brahma* and night of *Brahma* is of equal length. The age of *Manu, Manvantara* consists of seventy one times the age of the Gods.

A similar description of the four yugas occurs in the *Mahābhārata,* where, at the end of the cycle of four ages or the *Mahāyuga,* the *Kṛta* returns.[5] The emphasis here, as in *Manu,* is on the change in the character of the four ages accompanied by the increasing decline of *Dharma* in each, with a graphic picture of deterioration in the *Kali* age. At one point, it is stated that the world turns upside down in the *Kali* age.[6] It goes on to list the *Mlēccha* and other kings who will rule and the list seems to echo the more detailed lists of the *Purāṇas.* Parts of the *Mahābhārata* are regarded as later additions and some of these could well be contemporary with the early *Purāṇas* or at least were borrowing from a common source of ideas, current by the second or third centuries A.D.

The theory of the four ages as a conceptualization of Cosmological time is more elaborate in the *Purāṇas,* which are generally dated to around the mid-first millennium A.D. In the *Viṣṇu Purāṇa,* for instance, it is discussed in the sections on the creation of the world and concepts of Cosmology. The sentiment on the significance of *Dharma* remains evident, although there is a greater play with numbers in these texts.

The twelve thousand years are treated as divine years and conversion to human years requires multiplication by the hundred and sixty.[7] The length of the *Mahāyuga* is therefore calculated as 4,320,000 human years. The *Kalpa*, which incorporates the *Mahāyuga,* is calculated as one thousand ages or fourteen *Manvantaras*.[8] The *Manvantara* is equal to seventy one times the number of years in a *Mahāyuga* with some to spare. This we are told is equal to 852,000 (1200×71) divine years or 306, 720,000 (360×852,000) human years. It has been suggested that this is an imperfect synthesis of more than one independent doctrine and perhaps the *Manvantaras* derive from a source different to that of the *Mahāyugas*.[9] *Kali,* the smallest of the ages, extended to 432,000 years.

According to one view; the number 432,000 is of Babylonian origin and appears to have been combined with Greek epicycle theory.[10] The *Mahāyuga* was a category of time reminiscent of what geologists in the last century termed 'deep time' and perhaps what Fernand Brandel in our century would have recognized as 'la longu duree'—'the time of long duration'.

The projection of the *Kalpa* as Cosmological time in non-astronomical texts was almost calculated to defeat any controllable sense of time. Could the mathematical innovations of that period, namely, the concept of the Zero and of decimal place value notation have encouraged this formulation of figures and numbers? The magnification to millions was almost a fantasy on ciphers. There is perhaps something ironic in time constructs such as these playing with the concept of Zero which, as *Śūnya* could refer to emptiness or the void.

In the Cosmology of the *Purāṇas,* the *Kalpa* is the period through which creation lasts and is calculated in figures. The dissolution as the end of the *Kalpa* occurs when fire and flood engulf creation. In *Buddhist, Jaina* and *Ajīvika* Texts, the *Kalpa* was made unreal not just by the use of the fantasy of figure but also by a spatial description which made it virtual inconceivable. A *Buddhist* text describes the *Kalpa* thus :

There is a mountain in the shape of a cube, measuring one *Yōjana* and if every hundred years the mountain is brushed with a silk scarf, then the time that is taken for the mountain to be eroded by the scarf is the equivalent of a *Kalpa.*[11] Alternatively, if there is a city of iron walls, measuring one cubic *Yōjana,* and is filled with mustard seeds, and if one seed is taken out every hundred years, the seeds would all be removed before the *Kalpa* ended.[12] In *Jaina* texts, the wheel of time, *Kālacakra,* is said to consists of Zillions of atoms of time Palyopmas, calculated to 20×10,000,000 (2)×10×10,000,000. The measure of each is a gain described spatially; it is the number of years it would take to empty a cylinder, four miles in length and width, tightly packed with the body hair of humans, if every hundred years a single hair is taken out.[13] The *Ajīvika* description of a *Mahākalpa* states that if there was river 116,649 times (that is, seven to the seventh power) the size of the *Ganga,* and if every hundred years one grain of sand is removed from the bed of this imaginary river, the total time for the removal of all the sand will be one *Sara,* and 300,000 *Saras* equal one *Mahākalpa.* (Incidentally, it takes 8,400,000 of these to complete the transmigration of soul). This is not the total time of the Universe for it continues for infinitely longer.[14] Alternatively, space is also projected as time where the measurement of the Universe with a rope equals the distance which a male celestial being files in six months at the speed of 2,057,152 *Yōjanas* in one blinking of the eye.[15] In such reckonings a human life-time would constitute a minuscule moment, described in fact as a dew drop on the tip of a blade of grass when the Sun rises of the turning of the chariot wheel which turns by just one place on its rim.[16] Spatial descriptions of extended time are intended to suggest an infinity of time. But, since the silk scarf would have disintegrated in the first hundred years or so, spatial descriptions amount to a negation of measurement. It is, in this ultimate sense that the Mauryan emperor Aśōka refers to the continuation of his policies by his descendants and by posterity for a *Kalpa*.[17] An infinity of time comes to represent all, of creation.

In *Purāṇic* Cosmology, the *Mahāyuga* is more manageable than the *Kalpa* since it takes the figures from *Manu,* converts them to an equivalent in human years and the figures chosen continue to conform to a pattern. Each of the four *Yugas* decreases in regulated length with an intermediate dawn and dusk. The *Kṛta-Yuga* extends for 1,728,000 human years, which figure is divisible by 12. The *Trētā-Yuga* decreases in length by a quarter and is 1,296,000 years, and is divisible by 9. The *Dvāpara-Yuga* is reduced by a third and is 864,000 years in length and is divisible by 6. The *Kali-Yuga,* is reduced by a half and of 432,000 years in length and is divisible by 3.[18] The names of the four ages are called after the throw of the four-sided dice, from best of chance into what many otherwise appear to be pre-ordained. Play at dice is opposed to a permanent accumulation of wealth and has levelling effect sometimes even circulating that, which is scarce. Transactions of exchange tend to get neutralized as the game proceeds.[19] Was time seen as playing a similar role?

This construction of a time scale has many interesting facets. Orderliness of a mathematical kind was basic to the concept. The shortening of the length of each *Yuga* is in descending arithmetical progression. The numbers have a mathematical link as in figures such as 60,72,84,360, 1200, 432,000. The numbers also echo a variety of sources; 432,000 is said to be the number of the syllables in the *Ṛgvēda* and the number in the three *Vēdic* texts is double that, 864,000.[20] It is unlikely that a precise syllable count was made of the *Vēdas*. Were these numbers in turn picked up from other sources? In Mesopotamian Schedules, apart from the use of the basic sexi-gesimal system, the final sums are multiples of the same integer which in India came to be associated with the sum of divine years in a cosmic cycle. The multiplication of 1200 by 360 yields 432,000. This was also the time period given for the pre-Diluvian kings in Babylonian sources as recorded by Berossos, the Babylonian historian of the Third Century B.C.[21] During the previous centuries Achaemenid West Asia had encroached on North-Western India and possibly the number 72, which was basic to Biblical and Babylonian Chronology, became popular in Indian time calculations. From the Babylonian perspective too, some numerals such as 7, 9, 12, 360 were seen as endowed with a special magical potency and were therefore treated differently from figures in Mathematics.[22] Yet, these figures occur in Astronomy and the similarities may not be altogether an accidental. It would seem that such figures might have been travelling back and forth between India and West Asia as they continued to do into later centuries.

Further, in reckoning of time, it is suggested that each Yuga represents a progressive decline in plety, morality, strength, stature, longevity and happiness. This was constantly endorsed as characteristic of change over the four ages. Sometimes the description of the Yugas is merely a prelude to the lengthy statements on the reversal, from the *Brāhmaṇical* perspective, of norms mores, which reversal characterizes the *Kali* age. The gradual decline of *Dharma* is stated both directly and in some symbols. The utopian conditions of the first age, also sometimes referred to as *Satya,* the age of truth, diminish slowly until nothing of the Utopea is left in the *Kali Yuga.* Thus in the first age people lived 400 years, never suffering from disease or from insecurity. By contrast in the fourth age life expectancy comes down to 100 years and the *Śūdras* are in the ascendant.[23] In the *Kṛta* age men lived as long as they chose and procreation did not require sexual activity but in the *Kali* age marriage became necessary. In another text life expectancy drops from 30000 in the *Kṛta* age to 100 years in the *Kali* age. In the *Jaina* tradition the start of cycle was evident by the tallness of humans, a man's height being six miles and number of his ribs being 256. Gradually, both the height and number of ribs decreased until in the present era, life expectancy is down to 125, the number of ribs to 16 and the height to seven cubic hastas (about 18), and will decrease further in the future.[25] Or there is the image of *Dharma* sometimes likened to a bull, which stands on four legs in the first age and drops a leg in each subsequent age, symbolizing the decline in moral and social order.[26] The present is invariably an age of evil when social mores are turned upside down. This was in part the expected decline in the decreasing cycle where human actions would relate to the quality of the *Yuga.*

It could also be interpreted as resentment at royal patronage to non-*Brāhmaṇical* ideologies, for the *Brāhmaṇa* were not yet the forefront as recipients of material prosperity. The *Kali-Yuga* symbolizes the breaking down of caste ranking as a determining feature of social activities. Mention is made of *Mlēccha* rulers, corrupt *Brāhmaṇas* and upstrat *Śūdras* taking on the airs of *Brāhmaṇas* and performing priestly functions. Equally important, the subordination of women, so crucial to the continuance of caste, disintegrates and results particularly in sexual freedom for women.[27] This is unlikely as a description of an actual situation and the *Kali-Yuga* is here being construed as the antonym to an ideal *Brāhmaṇical* society. Historical evidence points to social change in the latter part of the first millennium A.D. but not of the kind envisaged in the description of the *Kali* age. For example, *Brāhmaṇa* ascending was established through extensive royal grants of land to a variety of *Brāhmaṇas* but this did not lead to a rescinding of the picture of the *Kali* age. The idyllic community of the distant past is expected to return but in the equally distant future. But the *Yugas* were not merely a measurement in time. Conflicting views among the authors of the texts setting out the norms could be sorted out by arguing that a particular interpretation, no longer current, was prevalent in the previous *Yuga*.[28] Substantial changes in customary laws and rituals explained as *Kalivarjaya* are in effect related to problems arising from social change in the latter part of the first millennium A.D.[29] The horrors of the *Kali Yuga* are recited repeatedly from this period onwards. One coming of the tenth *Avatāra* of *Viṣṇu, Kalkin*—a name also associated with time—will usher in a new age. *Kalkin* will destroy the *Mlēcchas,* the rules and heretics.[30] There is a torch of the messianic in the coming of *Kalkin*.

Slotted into this theory of the decline of *Dharma* is also the notion of transmigration or metapsychosis—*Karma* and *Saṃsāra*. The Cyclic notion is reinforced by the idea of transmigration with the *Upaniṣads* had tied together the notions of time, the planets and rebirth in the argument that after death the souls travel along two major paths—the *Dēvayāna* or path of Gods and the *Pitṛyāna* or path of ancestors. Those who were free from rebirth took the former path, which was associated with the Sun and led the soul to the world of *Brahma* and eternal life. Those who had to undergo rebirth took the path of the ancestors, associated with the Moon, and eventually returned to earth.[31] Implicit in this notion are certain consequences. There is the inevitability of death with which time comes to an end. Religious beliefs attempt to reverse this inevitability by proposing rebirth through many lives, or as in the semitic religions, everlasting life, heaven or hell subsequent to judgement day. In either case, one can't escape the consequences of one's actions. Therefore, if present action determines the future, then the past, present and future are inextricably linked.[32] This kind of thinking was in turn incorporated into horoscopy.[33]

The deterministic procession of time towards decline is arrested and challenged by the individual *Karma* and this idea became even more significant with the notion that *bhakti* (devotion) to and sharing in the grace of the deity—was an avenue of release from rebirth. It is at the human level that the immensity of time can be challenged for in its totality time is almost beyond human comprehension. There is a near dialectical relationship between the individuality of *Karma* and the virtual inevitability of the time cycle. The decline in the fourth age acts as a pen on which one can hand one's inability to change the course of events particularly if it points towards Tetrogression. But the four ages need not be perceived as enclosed units for it is said that a king's conduct characterizes the identity of the age and this ties ethics and social behaviour to time.[34] Ultimately the possibility of the return of the cycle provides the necessary optimism for continuing human action and also gives a meaning to human action in the past; it makes history necessary.

The change, implicit in each *Yuga* is so firmly emphasized that one may wonder how the idea that cyclic time did not conceive of change took root in modern views on the subject. Admittedly, in the existing *Mahāyuga*, the notion of change in the *Purāṇic* version is towards decline and not towards progress although the conduct of the king can change this identity. The linking of a particular period of time with social change marks difference between the *Purāṇic* Cosmological time with its implications for human society, and time as conceived in Astronomy. Jayasimha points to the decline of planetary motions as the *Mahāyuga* progresses, but this may be seen as a different concept of decline from the *Purāṇic*. The figures from Astronomy are relatively value-free, arising out of computations of time. This difference is important to the relationship between time and history.

The induction of Cosmological time into the early *Purāṇas* raises the question of the relationship of myths to history. It has been argued that myths narrate events in primordial, atemporal moment which constitute sacred time and differ from the continuous profane time of daily routines. Therefore, by narrating myths profane time is abolished.[35] The question then is whether this was true of the treatment of time in the *Purāṇas*. Although not stated directly, was there at least the suggestion that a differentiation was being perceived between different forms of time and the assessment of past events? Was mythic time segregated from chronological time in narrating the past? Was this segregation seen as a mechanism for separating myth and history?

Among the Purāṇas, the *Viṣṇu Purāṇa* observes what is described as the ideal format in its five sections and one of the five sections provides an overview of the perceived past. This is the *Vaṃśānucarita* which narrates the succession of those who ruled and can be divided into the three distinct, although unequal, sections.[36] The narrative begins with the briefest reference back to the previous sections of the text which describe the genesis of the Universe, giving details of creation, of time cycles as both *Mahāyugas* and the *Manvantaras,* and of Cosmography. The Manus are mentioned in passing, spanning immense cycles of time. They have no association with specific events nor do they indicate any particular social condition.

In the reign of the seventh Manu there occurs the great flood from which Manu is saved by *Viṣṇu* in his fish incarnation, the *Matsya-Avatāra*. Manu and the seven *Ṛṣis* are placed in a boat and the boat is tied to the single horn of the fish (the incarnated *Viṣṇu)* and they go through the waters of the flood and are lodged safely on the top of a mountain (in some texts, this is Mount *Mēru,* the axis *Muṇḍi*). When the flood subsides, creation starts anew. The story of the flood is in many ways to similar to the Mesopotamian story. The earliest version occurs in the *Śatapatha Brāhmaṇa* and may reflect a borrowing from West Asia where there are still earlier versions of the story.[37] The flood acts as a time marker separating the period of the Manus from that of the *Kṣatriya Rājas* who followed. Floods have a dual symbolism of water washing away that which is physically present as well as the silt deposited from the flood providing a fresh beginning in the space which had been flooded.

In the next and next second section of the chapter on succession, the measurement of time changes radically from the cyclic time associated with the Manus and the *Manvantaras,* to reckoning time as generations through lengthy genealogies. So both the flood and the change to reckoning in generational time through genealogies are mechanisms of demarcating myth in Cosmological time from that which occupies the ambiguous area approaching history.[38]

The listing of generations in the *Viṣṇu Purāṇa* begins

with description of Manu's progeny who are said to be the ancestors of the earliest *Kṣatriyas.* Among them the most important are *Ikṣvāku* who initiates the *Sūryavaṃśa* or Solar lineage, and *Iḍa,* whose androgynous form is suggested and from whose female self are descended another group of *Kṣatriyas* identified as the *Candravaṃśa* or Lunar lineage. These two descent groups are also integrated into the two epics, for the heoes of the *Rāmāyaṇa* are of the *Sūryavaṃśa* and those of the *Mahābhārata* are of the *Candravaṃśa*. The association of the Sun and the Moon with these descent groups carries a rich symbolism which touches on many facets but for our purpose here it is, significant as a parallel to the calendars in use. Time reckoning in this section is characterised by the sequence of generations and is for all practical purposes and exercise in linear time, although a location within the *Yuga* system forms the ultimate time frame.

The term used for a succession list of any kind or a genealogy is *Vaṃśa,* derived from the name for bamboo. This is an appropriate image where each node marks a new generation of growth. This is in itself would suggest linear time. Genealogies have tentative beginnings in the early chapters of the epics, but the construction of a narrative in genealogical form comes into its own in the early *Purāṇas.* This may be called a construction since it covers almost a hundred generations and such an immense genealogy could hardly have been on authentic record.

The genealogies in themselves are not invariably authentic but they incorporate patterns and narratives which are indicative of recognition of particularities. For example, the *Sūryavaṃśa* pattern is the history of descent from father to eldest son alone, generally suggestive of primogeniture, whereas the *Candravaṃśa* often lists more than a single time of descent among brothers, suggestive perhaps of a segmentary lineage system. The genealogies are therefore not necessarily intended as historically accurate—and this would be evident from the variants in the name and position in the descent list among the different *Purāṇas*—but they do seem to be making a statement about how the past was perceived and these perceptions may be intended as the more authentic history.

The descent list of the *Sūryavaṃśa* peters out. The *Candravaṃśa,* purporting to record segmentary clans, continues to the grand finale of the *Mahābhārata* war in which virtually all the clans are directly or indirectly involved and few survive. The catastrophic war, described as the end of the glorious age of the *Kṣatriyas,* becomes another time-marker, if the choice of a time marker were a signpost to the culture, it would seem appropriate that *Kṣatriya* clans would terminate their activities in a war. The keeping of genealogies and reckoning by generations, whether factual or fictive, are important to lineage-based societies where rank is determined by birth and such clans claim dominant social status.[39] Genealogical time becomes important to the process of legitimation. Earlier generations can be stretched back into a remote past or claims to heroic ancestry, can be telescoped so to be placed closer to the claimant.

That, both the flood and the *Mahābhārata* war were viewed as significant time-markers is indicated by the uncertainty as to where the start of the *Kaliyuga* should be placed.[40] The *Viṣṇu Purāṇa* states that the *Kali* age began after the wars,[41] although some other texts place it earlier, thus eliminating a precise time. However, the start of the *Kaliyuga* in the *Viṣṇu Purāṇa* is located with reference to a constellation, thus implicitly suggesting a point in time. The astronomers, for reasons of facilitating calculations, worked out the date of the *Kaliyuga* as equivalent to 3102-1 B.C. This was the date which came to be adopted in inscriptions and histories as well where precise dates were required.[42] But, in the *Purāṇas* it was viewed more as the condition of society than as a precise point in time. Cosmological time did not debar other forms of time reckoning, perhaps more realistic and therefore liable to be subsumed in Cosmological time.

Subsequent to the *Mahābhārata* war, in the third section of the chapter on succession in the *Viṣṇu Purāṇa,* the narrative dramatically changes its tense from the past to the future and becomes a lengthy prediction of events. This would seem to be another example of pointing to historical change, through adopting a new category of time. What follows soon after is a listing of dynasties which is a change from the earlier lists of descent groups and within these dynasties the succession of rulers with, in some *Purāṇas,* their regnal years. Time periods and regnal years of the initial dynasties are obviously exaggerated but the lists gradually became more credible. The genealogical form continues and the succession of dynasties as a new pattern of chronology is suggestive of linear time. The genealogical form is however limited to the names within the dynasty. Dynasties as such are generally not released to each other in some cases it is specifically mentioned that the successor dynasty began as a *Bhṛtya* or in the service of the earlier one. The association with the *Kaliyuga* is less in terms of a point of time and more as an explanation of why, when the *Kṣatriya* clans died out, it was possible not only for *Śūdras* to rule—which at that time was contrary to the norms-but it was also possible later for kings to be formed socially unacceptable groups, such as the *Mlēcchas* or those outside the social pale, the *Vratya-Kṣatriyas* or degenerate *Kṣatriyas* and the *Śabaras* and *Pulindas* who were forest dwellers and therefore regarded as primitive and beyond the boundaries of caste society.

Prophesying history becomes a mechanism for using the past to lay claim to controlling the future. But the use of the future tense had its obvious limitation, namely that the prophecy had to stop at the point when the text was composed, in this case towards the mid-first millennium A.D.

Within the narrative of the chapter on succession in the *Viṣṇu Purāṇa* there is a demarcation of three periods based on various categories of time. The first is remote, that of the Manus and their large time cycles; the second has faint continuity from the first but focuses on a changed chronological pattern through genealogical or generational time; and finally, subsequent to the *Mahābhārata* war there follows a period of dynastic time reckoning. Cosmological time in this context, although all encompassing, is nevertheless viewed as distant and possibly as mythical time. Dynastic time in effect takes the functional form of historical chronology. A distinction between myth and history, although, not stated in these terms, appears to have been perceived. The coming of the *Kaliyuga* heralds a major change in time reckoning. The forms in which time is depicted seem to be making a statement about historical change.

Historical Time

A more realistic kind of time reckoning can be drawn on actual historical needs. Cosmological time didn't debar other forms of time reckoning, even those perhaps with shorter spans and therefore liable to be subsumed in the compass of Cosmological time. These may be viewed as fragmentary arcs within the cycle, which take on the role of linear time. The dichotomy between cyclic and linear becomes increasingly vague.

An innovation of a new kind drawing on both Astronomy and History provided another set of time markers and inaugurated a system of time reckoning which was distinctively different. This has curiously been overlooked by those insisting on early Indian time concepts being exclusively cyclic. The use of eras as well as dating by regnal years had gained currency, simultaneously with activity in Astronomy and Cosmology. This is particularly evident in documents concerning those in political authority. Regnal years go back to the earliest historical inscriptions, the edicts of the Mauryan Emperor Aśōka, issued in the third century B.C. Regnal years and eras point to time reckoning being based on a precise point in time and the starting point of the reckoning being known.

Linear time came to be used more extensively from the Christian era onwards in variety of ways. Dynastic lists gradually came into circulation. Biographies of those in power, and especially kings, took as their time bracket the dynasty to which the subject of the biography belonged, relating him both to a description of origins and to early rulers, but focussed on the major activities of particular kings. The use of historical eras located kings and dynasties more firmly in time. The event or the person became a point in time and became stable within that tradition. The use of eras gained currency and this is particularly evident in documents emanating from those in authority.

The systems of eras, according to some scholars grew out of points in time invented to facilitate calculations in Astronomy. The *Kṛta era* of 58 B.C., also known as the *Māḷava era,* and later, from the eight century A.D. referred to as the *Vikrama era,* has been claimed as commemorating a historical event, although the suggestion has also been made that it may be associated with Astronomy.[44] The term *Saṃvatsara* came to be used for era. Other commonly used eras were the *Śaka* (A.D. 78), the *Kālacuri Cēdi* (A.D.

247-8), the *Gupta* (A.D. 319-20) and so on. Subsequent to this there is a mushrooming of new eras all over the sub-continent.

The historical use of an era sometimes associated it with either an accession—such as the *Gupta era*—or with being status symbol adding lustre to a reign—as in the case of the *Cālukya-Vikrama era.* At the functional level, eras introduced more precise dating in official documents. But this was not the only significance of using an era. More importantly, it helped separate Cosmological time reckoning from the functional, although the former was not discarded. That the fashion for and computation of eras began from the use of methods influenced by Hellenistic Astronomy may be one explanation, but it was also related to other changes, which were conducive to the establishing of eras.

Dynasties are associated with monarchies and these in turn pre-suppose the emergence of a state system. The adoption of a short-spanned and more precise time reckoning associated with those in authority was encouraged by the establishing of a state. This time reckoning, even when encompassed by the longer spans of Cosmological time, introduces a precision, which sends strength to the authority of the State. The need to maintain state records and official documents, not to mention, texts legitimizing persons and institutions, tended to make the history of those in authority more evident. The time dimension of these was effectively linear. Elements of linear time in the context of early India have, therefore, a different origin from linear time in the Judaeo-Christian tradition, where it is linked to the sacred.

The use of historical eras may have been partially influenced by the parallel tradition of the Buddhists where major events came to be dated in the number of years from the *Mahāpari Nirvāṇa* or the passing away of the *Buddha.*

The date for this, it was believed, was carefully recorded and after the removal of some discrepancies in late texts, most scholars, basing themselves on sources in question; accepted 486 B.C. or 483 B.C. as the date of the death of the *Buddha*. Recently this date has been questioned and it is being suggested that the *Buddha* may have lived and died almost a century later.[45] What is significant, however, is that within particular Buddhist traditions there was stable date. Perhaps because of this stable date, chronicles written by Buddhist monks, for example, display a sharper sense of time and relate all events to this date, which is associated with a historical person. Perhaps the historicity of the *Buddha* and the establishing of the *Sangha* separated myth from history, although the narrative of individual events was often wrapped in myths. *Buddhism* maintained monastic institutions, the chronology and history of which were in some cases recorded when they became centres of power. Sectarian break aways, of which there were many, could claim greater legitimacy if they placed themselves in relation to this history. Proselytizing ideologies claim legitimacy, both from myth and resumed history. There was also, for instance in the Sri Lankan Chronicles, such as *Dīpavaṃśa* and the *Mahāvaṃśa,* a reconstruction of regnal years and events relating to the earliest narrative about the history of Sri Lanka, in order to correlate events in Sri Lanka with the biography of the Buddha. Later the history of the *Sangha* was integrated with the political history of Sri Lanka and in its initial stages even with the rule of the Mauryas in India.[46] This became necessary because Aśōka came to be projected as legitimizing the sect of the *Thēravāda* and the authors of these chronicles were *Thēravāda* monks.

The adoption of historical eras or regnal years meant that official documents began to carry dates. Among these documents are inscriptions that are in effecting the annals of Indian history. Issued by individual rulers and by members of the ruling class in the main, they include votive records relating to grants and gifts as well as statements of events, especially those important to politics and administration. Some carry brief histories of the dynasty as a prelude to the specific action recorded. The date is precise and includes the era of the regnal year of the ruler, the season, month, fortnight and the day. It follows calendrical requirements for a date and this enables the calculation, for example, of its equivalence in the Gregorian calendar. Facility in the use of calendars would have required professional expertise. Hence, the continuing importance of the astrologer, who, apart from horoscopy, sustained his authority by keeping track of both Lunar and Solar calendars.[47] The status of the astrologer in the earlier texts remained low even within the *Brāhmaṇa Varṇa* and he was excluded from *Śrāddha* ceremonies but interestingly, the later *Bṛhatsaṃhita,* endorsing Astrology, contradicts this and refers to the astrologer as the *Saṃvatsara* and requires that he be honoured as the chief guest at a *Śrāddha*.[48]

Precision in dating inscriptional documents became necessary because they were often official statements or legal charters. The recording of a precise moment requires a context and location. Alternatively, depending on the nature of the grant, it may have been required to negate the effect of something inauspicious such as an eclipse. In such cases the precision in dating had to do with an astrologically appropriate moment. Despite the interest in horoscopes, history became more recognizable and in its use of time reckoning, it gradually slashed off some of its earlier ritual context.

Apart from inscriptions, other categories of records such as biographies and dynastic regional chronicles, the *Caritas* and *Vaṃśāvalis*—articulated views on the past. These evolved into genres of literature and borrowed from the country literary style. But their functions, as records and as legitimizing those in power were new and significant. Biographies, particularly of kings, took as their time bracket the dynasty to which the subject of the biography belonged. They focussed on what they assessed, as the more significant events of the reign where authority is specific, the recording of time also tends to be more specific. The composition and records of poets and bards, sometimes close to the court, but often closer to popular sentiments, were yet another form of legitimization, but could occasionally context the official version of events.[49]

The *Caritas* and *Vaṃśāvalis* drew, when they choose to, on the *Purāṇas* but their view of the past was not identical to that of the *Purāṇas*.[50] Examples of such *Caritas* are the *Harṣacarita* of *Bāṇabhaṭṭa,* the biography of the Seventh Century ruler *Harṣavardhana,* or *Bilahaṇa's Vikramāṅka-dēvacarita* which focuses on *Vikramāditya-VI,* an eleventh century *Cālukya* king.[51] Some of the lengthier inscriptions carried narratives of the dynasty and among these in the tenth century Khajuraho inscription set up by *Dhanga* for his father *Yaśōvarman*.[52] This is an official version of the origins and early history of the *Cāndella* dynasty. Two contested versions are known : one refers to their origins and caste status which is placed low and the other is an alternative view of the reign of one of the kings from the perspective of a popular ballad. An example of the more common form of the *Vaṃśāvali* is the one from *Chamba*.

These texts and inscriptions pertaining to past events frequently began with the origin myths of the dynasty as a preamble. Origin myths could go back to the time cycles of the Manus and related to non-calendrical time but could also be set in the genealogical section of what may be called generational time.[53] This serves to legitimize the founders of a dynasty by linking them with the heroes of *Purāṇic* ancestry. Some, however, may not resort to *Purāṇic* ancestry, although it tends to be brought in indirectly. By way of contrast, the narrative, as it continues in such texts, relates the succession of later rulers. This is set in a matrix of generations and of dynastic lists, emphasizes the unique event, and goes beyond the *Purāṇic* view of the past in, what could be suggested were the major concerns perceived as important to historical change. These were primarily the recognition of the importance of acquiring caste status, the emergence of institutions linked to state formation and the establishing of new religious sects supportive of monarchy.[54] Thus in the *Chamba Vaṃśāvali,* the origin myth merely links the earliest time with the heroes of *Purāṇic* ancestry. The subsequent narrative rulers of *Chamba* revolve around change of many kinds, such as the claims to high status, the shifting of the capital to an area conducive to revenue appropriation and administrative control, the making of grants, the introduction of *Vaiṣṇavism,* and so on. The chronology is what we would recognize as historical categories of time demarcate the earlier story from the later narrative. Such devices of temporal distancing also provide of how the past was perceived.

A refutation of history through recourse to cyclic time was identified more frequently with colonial cultures. This might explain the reluctance and even the dismissal of attempts to inquire into how these concepts of time were constructed, after the failure of the initial attempts to comprehend them. H.H. Wilson, among the more respected Indologists of the 19th century, commenting on the description of time in the *Viṣṇu Purāṇa,* makes a telling statement. "It does not seem necessary to refer the invention (of time) to any astronomical computations or to any attempt to represent actual chronology".[55] Had the initial enquiry been pursued, the link between astronomical computations and the grand design of cyclic time as well as the more mundane time reckoning of historical chronology would have become more visible. The standing away of myth from history was also sometimes articulated in categories of time. The simultaneous use of different categories of time is symbolic of registering historical change and priorities in historical functioning. These differentiations, seem to be significant, not because in the 20th century we are analyzing past concepts of time, but because there was a reason in the past for making these differentiations.

The characterizing of societies as using either cyclic or linear time is an inadequate explanation for the centrality or otherwise of history. Research even into European history, has endorsed various categories of historical time and to that extent has distanced itself from the earlier single category. Time, as conceived in Cosmology does not exclude the use of other categories of time and these can be simultaneous in the same society. It seems more appropriate to enquire into how a society uses a particular category and what is being intended by that use. Thus, the statements which form of time reckoning are not identical. The inclusion of cyclic time is not a characteristic of cultures, which are historically stunted, but an indication of historical complexity. This complexity is reflected in the perceptions of the past in pre-modern times the premises of which were different from the writing of history today.

REFERENCES

1. यद्वेवैनँ शिक्येन बिभर्ति। संवत्सर एषोग्निर्ऋतवः शिक्यमृतुभिर्हि संव्वत्सरः शक्नोति स्थातुं यच्छक्नोति तस्माच्छिक्यमृतुभिरेवैनमेतद्बिभर्त्ति षड्डुद्यामं भवति षड्ढचृतवः॥

शतपथब्राह्मणम्, VI.७.१.१८.

Yadvēvainaṁ Śikyēna Bibharti। Saṃvatsara Ēṣōgnirṛtavaḥ Śikyamṛtubhirhi Saṃvvatsaraḥ Śaknōti Sthātuṃ Yacchaknōti Tasmācchikyam ṛtubhirēvainam ētadbibhartti Ṣaḍudyāmaṃ Bhavati Ṣaḍḍhacṛtavaḥ॥

Śatapathabrāhmaṇa, VI.7.1.18.

तस्यायमेव लोकः प्रथममहः। अयमस्य लोको व्वसन्तऋतुर्य्यदूर्ध्वमस्माल्लोकादर्व्वाचीनम् अन्तरिक्षात्तद्द्वितीयमहस्तद्वस्य ग्रीष्मऋतुरन्तरिक्षमेवास्य मध्यममहरन्तरिक्षमस्य व्वर्षाशग्दावृतू यदूर्ध्वमन्तरिक्षादर्व्वाचीनन्दिवस्तच्चतुर्थमहस्तद्वस्य हेमन्तऋतुर्द्यौरेवास्य पञ्चममहर्द्यौरस्य शिशिरऋतुरित्यधिदेवतम्।

शतपथब्राह्मणम्, XIII.६.१.१०.

Tasyāyamēva Lōkaḥ Prathamamahaḥ। Ayamasya Lōkō Vvasanta Ṛtuḥ Yadūrdhvamasmāllōkādarvvācīnam Antarikṣāt Tat Dvitīyamahastadvasya Grīṣmarturantarikṣamēvāsya Madhyamamaharantarikṣamasya Vvarṣāśagdāvṛtū Yadūrdhvamantarikṣādarvvācīnandivastaccaturthamahaḥ Tadvasya Hēmanta Ṛturdyaurēvāsya Pañcamamahardyaurasya Śiśirarturityadhidēvatam।

Śatapathabrāhmaṇa, XIII.6.1.10.

अथाध्यात्मम्। प्प्रतिष्ठैवास्य प्प्रथममहः प्रतिष्ठो अस्यव्वसन्तऋतुर्य्यदूर्ध्वम्प्रतिष्ठाया अवाचीनम् मध्यात्तद्द्वितीयमहस्तद्वस्य ग्ग्रीष्मऋतुर्म्मध्यमेवास्य मध्यममहर्म्मध्यमस्य व्वर्षाशरदावृतू यदूर्ध्वं मध्यादवाचीनँ शीर्ष्णस्तच्चतुर्थ्थमहस्तद्वस्य हेमन्तऋतुः शिरएवास्य पञ्चममहः शिरोस्य शिशिरऋतुरेवमिमे च लोकाः सम्वत्सरश्चात्मा च पुरुषमेधम- भिसम्पद्यन्ते सर्व्वम्वाइमे लोकाः सर्व्वं सम्वत्सरः सर्व्वमात्मा सर्व्वम्पुरुषमेधः सर्व्वस्याप्त्यै सर्व्वस्यावरुद्धचै॥११॥

Athādhyātmam। Ppratiṣṭhaivāsya Pprathamamahaḥ Pratiṣṭhō Asyavvasanta Ṛturyyadūrdhvampratiṣṭhāyā Avācīnam Madhyāttaddvitīyamahastadvasya Ggrīṣma Ṛtuḥ Madhyamēvāsya Madhyamamahaḥ Madhyamasya Vvarṣāśaradāvṛtū Yadūrdhvaṃ Madhyādavācīnaṁ Śīrṣṇastaccaturththamahaḥ Tadvasya Hēmantarṭuḥ Śiraēvāsya Pañcamamahaḥ Śirōsya Śiśirarturēvamimē Ca Lōkāḥ Samvatsaraścātmā Ca Puruṣamēdhamabhisampadyantē Sarvvamvāimē Lōkāḥ Sarvvaṁ Samvatsaraḥ Sarvvamātmā Sarvvam Puruṣ amēdhaḥ Sarvvasyāptyai Sarvvasyāvaruddhacai।।11।।

स वै व्वौगिति करोति व्वाग्वै व्वषट्कारो वाग्रेतो रेत एवैतस्तिञ्चति षडित्यृतवो वै षट् तदृतुष्वेवैतद्रेतः सिच्यते तदृतवो रेतः सिक्तमिमाः प्रजाः प्रजनयन्ति तस्मादेवं व्वषट्करोति॥

शतपथब्राह्मणम्, १.७.२.२१.

Sa Vai Vvaugiti Karōti Vvāgvai Vvaṣaṭkārō Vāgrētō Rēta Ēvaitastiñcati Ṣaḍityṛtavō Vai Ṣaṭ Tadṛtuṣvēvaitadrētaḥ Sicyatē Tadṛtavō Rētaḥ Siktamimāḥ Prajāḥ Prajanayanti Tasmādēvaṃ Vvaṣaṭkarōti।।

Śatapathabrāhmaṇa, I.7.2.21.

अथ नीविमुद्वह्य नमस्करोति। पितृदेवत्या वै नीविस्तस्मान्नीविमुद्वह्य नमस्करोति यज्ञो वै नमो यज्ञियानेवैनानेतत्करोति षट् कृत्वो नमस्करोति षड्वा ऋतव

ऋतवः पितरस्तस्मात्षट्कृत्वो नमस्करोति गृहान्नः पितरो दत्तेति गृहाणाँ ह पितर ईशत एषो एतस्याशीः कर्मणोथावजिघ्रति प्रत्यवधाय पिण्डान्त्स यजमानभागोग्नौ सकृदाच्छिन्नान्यभ्यादधाति पुनरुल्मुकमपिसृजति॥

शतपथब्राह्मणम्, II.४.२.२४.

Atha Nīvimudvṛhya Namaskarōti | Pitṛdēvatyā Vai Nīvistasmānnīvimudṛhya Namaskarōti Yajñō Vai Namō Yajñiyānēvainānētatkarōti Ṣaṭ Kṛtvō Namaskarōti Ṣaḍvā Ṛtava Ṛtavaḥ Pitarastasmātṣaṭkṛtvō Namaskarōti Gṛhānnaḥ Pitarō Dattēti Gṛhāṇām Ha Pitara Īśata Ēṣō Ētasyāśīḥ karmaṇōthāvajighrati Pratyavadhāya Piṇḍāntsa Yajamānabhāgaḥ Agnau Sakṛdācchinnānyabhyādadhāti Punarulmukamapisṛjati ||

Śatapathabrāhmaṇa, II.4.2.24.

ग्रीष्मो हेमन्तः शिशिरो वसन्तः शरद् वर्षाः स्विते नो दधात।
अ नो गोषु भजता प्रजायां निवात इद् वः शरणे स्याम॥

अथर्ववेद, ६.५५.२.

Grīṣmō Hēmantaḥ Śiśirō Vasantaḥ Śarad Varṣāḥ Svitē Nō Dadhāta |
A Nō Gōṣu Bhajatā Prajāyāṃ Nivāta Id Vaḥ Śaraṇē Syāma ||

Atharvavēda, VI.55.2.

2. सप्त युञ्जन्ति रथमेकचक्रमेको अश्वो वहति सप्तनामा।
त्रिनाभि चक्रमजरमनर्वं यत्रेमा विश्वा भुवनाधि तस्थुः॥

ऋग्वेद, १.१६४.२.

Sapta Yuñjanti Rathamēkacakramēkō Aśvō Vahati Saptanāmā |
Trinābhi Cakramajaramanarvaṃ Yatrēmā Viśvā Bhuvanādhi Tasthuḥ |

Ṛgvēda, I.164.2.

यत् पुरुषेण हविषा देवा यज्ञमतन्वत।
वसन्तो अस्यासीदाज्यं ग्रीष्म इध्मः शरद्धविः॥

ऋग्वेद, X.९०.६.

Yat Puruṣēṇa Haviṣā Dēvā Yajñamatanvata |
Vasantō Asyāsīdājyaṃ Grīṣma Idhmaḥ Śaraddhaviḥ ||

Ṛgvēda, X.90.6.

वसिष्ठं ह वरुणो नाव्याधादृषिं चकार स्वपा महोभिः।
स्तोतारं विप्रः सुदिनत्वे अह्नां यान्नु द्यावस्ततनन् यादुषासः॥

ऋग्वेद, ६.८८.४.

Vasiṣṭhaṃ Ha Varuṇō Nāvyādhādṛṣiṃ Cakāra Svapā Mahōbhiḥ |
Stōtāraṃ Vipraḥ Sudinatvē Ahnāṃ Yānnu Dyāvastatanan Yāduṣ āsaḥ ||

Ṛgvēda, VII.88.4.

3. Ṛgvēda, I, 164. 13-34.
4. Manu, I, 60-86.
5. Vanaparvan, 186, 17 ff; Śāntiparvan, 224, 6 ff.
6. Vanaparvan, 186, 28 ff.
7. दिव्यैर्वर्षसहस्त्रैस्तु कृतत्रेतादिसंज्ञितम्।।
   चतुर्युगं द्वादशभिस्तद्धिभागं निबोध मे।।

   विष्णु पुराण, १.३.११ ff.

   Divyairvarṣasahasraistu Kṛtatrētādisañjñitam//
   Caturyugaṃ Dvādaśabhistaddhibhāgaṃ Nibōdha Mē// .....

   Viṣṇu Purāṇa, 1.3.11 ff.
8. *Ibid.*, III, 2.
9. A.L. Basham, 'The Wonder that was India', London, 1954, p. 321.
10. *Ibid.*
11. Samyukta Nikāya, II, 180-1.
12. *Ibid.*, 181-2.
13. W. Norman Brown, 'Man, in the Universe', Berkeley, 1966, p. 77.
14. A.L. Basham, 'History and Doctrine of the Ajivikas', London, 1951, p. 253-4.
15. A Ghosh (ed) 'Jaina Art and Architecture', 3 Vols. New Delhi, 1974-5, 3, 519, n 2
16. Visuddhimagga, I, 231, 238, quoted in A.K. Kumaraswamy 'Time and Eternity', Ascona, 1947.
17. Fourth Major Rock Edict. J. Bloch, Les inscriptions d' Aśōka, 1950, 100; Fifth Major Rock Edict, *Ibid.*, 102.
18. Viṣṇu Purāṇa I, 3. 11 ff; VI, 1;
   चत्वारि भारते वर्षे युगानि ऋषयोब्रुवन्।
   कृतं त्रेता द्वापरं च कलिश्चैवं चतुर्युगम्।।

   मत्स्य पुराणम्, १४२.१७.

   Catvāri Bhāratē Varṣē Yugāni Ṛṣayōbruvan/
   Kṛtaṃ Trētā Dvāparaṃ Ca Kaliścaivaṃ Caturyugam//

   Matsya Purāṇa, 142.17.

   पूर्वं कुतयुगं नाम ततस्त्रेतभिधीयते।
   द्वापरं च कलिश्चैव युगानि परिकल्पयेत्।।१८।।
   Pūrvaṃ Kutayugaṃ Nāma Tatastrētabhidhīyatē/
   Dvāparaṃ Ca Kaliścaiva Yugāni Parikalpayēt//18//
19. *Ibid.*
20. P.V. Kane 5.1. 689-90;
   स ऐक्षत प्रजापतिः। त्रय्यां वाव व्विद्ययाँ सर्व्वाणि भूतानि हन्त त्रयीमेव व्विद्यामात्मानमभिसँस्करवा इति।।२२।।

   शतपथब्राह्मणम्, १०.४.२.२२.

Sa Aikṣata Prajāpatiḥ। Trayyāṃ Vāva Vvidhyāyāṁ Sarvvāṇi Bhūtāni Hanta Trayīmēva Vvidyāmātmānamabhisaṁskaravā iti।।22।।

Śatapathabrāhmaṇa, X.4.2.22.

स ऋचो व्व्यौहत्। द्वादश बृहतीसहस्राण्येतावत्यो हर्चो याः प्रजापतिसृष्टास्तास्त्रिंशत्तमे व्व्यूहे पङ्क्ष्विवतिष्ठन्त ता यत्त्रिंशत्तमे व्व्यूहेतिष्ठन्त तस्मात्त्रिंशन्मासस्य रात्रयोथ यत्पङ्क्तिषु तस्मात्पाङ्क्तः प्रजापतिस्ता अष्टाशतँ शतानि पङ्क्त्योभवन्॥२३॥

Sa Ṛcō Vvyauhat। Dvādaśa Bṛhtīsahasrāṇyētāvatyō Harcō Yāḥ Prajāpatisṛṣṭāstāstriṁśattamē Vvyūhē Paṅktiṣvatiṣṭhanta Tā Yattriṁśattamē Vvyūhētiṣṭhanta Tasmāttriṁśanmāsasya Rātrayōtha Yatpaṅktiṣu Tasmātpāṅktaḥ Prajāpatistā Aṣṭāśataṁ Śatāni Paṅktayōbhavan।।23।।

अथेतरौ व्वेदौ व्व्यौहत्। द्वादशैव बृहतीसहस्राण्यष्टौ यजुषां चत्वारि साम्नामेतावद्धैतयोर्व्वेदयोर्यत्प्रजापतिसृष्टं तौ त्रिंशत्तमे व्व्यूहे पङ्क्तिष्वतिष्ठेतां तौ यत्त्रिंशत्तमे व्व्यूहेतिष्ठेतां तस्मात्त्रिंशन्मासस्य रात्रयोथ यत्पङ्क्तिषु तस्मात्पाङ्क्तः प्रजापतिस्ता अष्टाशतमेव शतानि पङ्क्त्योभवन्॥२४॥

Athētarau Vvēdau Vvyauhat। Dvādaśaiva Bṛhatīsahasrāṇyaṣṭau Yajuṣāṃ Catvāri Sāmnāmētāvaddhaitayōrvvēdayōryatprajāpatisṛṣṭaṃ Tau Triṃśattamē Vvyūhē Paṅktiṣvatiṣṭhētāṃ Tau Yattriṁśattamē Vvyūhētiṣṭhētāṃ Tasmāt Triṁśanmāsasya Rātrayōtha Yatpaṅktiṣu Tasmātpāṅktaḥ Prajāpatistā Aṣṭāśatamēva Śatāni Paṅktayōbhavan।।24।।

ते सर्व्वे त्रयो व्वेदाः। दश च सहस्राण्यष्टौ च शतान्यशीतीनामभनन्त्स मुहूर्तेन मुहूर्तेनाशीतिमाप्नोन्मुहूर्तेन मुहूर्तेनाशीतिः समपद्यत॥२५॥

Tē Sarvvē Trayō Vvēdāḥ। Daśa Ca Sahasrāṇyaṣṭau Ca Śatānyaśītīnāma-Bhanantsa Muhūrtēna Muhūrtēnāśītimāpnōnmuhūrtēna Muhūrtēnāśītiḥ Samapadyata।।25।।

21. *Ibid.*
22. *Ibid.*
23. Manu, I, 83; Kane, 32, 44 ff, 892 ff; Viṣṇu Purāṇa, 61 & 2; Mahābhārata Vanaparvan, 148, 10 ff; 186, 23 ff.
24. Mahābhārata, Śāntiparvan, 200, 35 ff.
25. S. Stevenson, 'The Heart of Jainism', New Delhi, 1970.
26. Manu, I, 81-2; VIII, 16.
27. Viṣṇupurāṇa, VI, 1.10 ff; Mahābhārata, Vanaparvan, 188.10-93.
28. P.V. Kane, 'History of Dharmaśāstra', Vol. III, Bandarkar Oriental Research Institute, 1946, p. 865 ff.

29. *Ibid.*, p. 926 ff, p. 966 ff.
30. Viṣṇu Purāṇa, IV, 24. 98; P.V. Kane, III, p. 923 ff.
31. तद्विद्वाँसः श्रोत्रियाः। अशिष्यन्त आचामन्त्यशित्वाचामन्त्येतमेव तदनमनग्नं कुर्व्वन्तो मन्यन्ते तस्मादेवम्विदशिष्यन्नाचामे दशित्वाचामेदेतमेव तदनमनग्नङ्कुरुते॥
बृहदारण्यकोपनिषद्, ६.२.१५.
Tadvidvāṁsaḥ Śrōtriyāḥ । Aśiṣyanta Ācāmanti Aśitvācāmantyētamēva Tadanamanagnaṃ Kurvvantō Manyantē Tasmādēvamvidaśiṣ yannācāmē Daśitvācāmēdētamēva Tadanamanagnaṅkurutē ।।
Bṛhadāraṇyakōpaniṣad, VI.2.15.
त द्य इत्थं विदुः ये चेमेरण्ये श्रद्धा तप इत्युपासते तेर्चिष मभिसम्भवंत्यर्चिषोहरह्न आपूर्यमाणपक्षमापूर्यमाणपक्षाद्यान् षडुदङ्ङेति मासां स्तान्॥१॥
छांदोग्योपनिषद्, ५.१०.१.
Ta Dya Itthaṃ Vidhuḥ Yē Cēmēraṇyē Śraddhā Tapa Ityupāsatē Tērciṣa Mabhisambhavantyarciṣōharahna Āpūryamāṇapakṣ amāpūryamāṇapakṣādyān Ṣaḍudajñēti Māsāṃ Stān ।।1।।
Chāndōgyōpaniṣad, V.10.1.
मासेभ्यस्संवत्सरं संवत्सरा दादित्य मादित्याच्चन्द्रमसं चन्द्रमसो विद्युतं तत्पुरुषोमानवस्स एना न्ब्रह्मगमयत्येष देवयानः पन्था इति॥२॥
Māsēbhya Ssaṃvatsaraṃ Saṃvatsarā Dāditya Mādityāccandramasaṃ Candramasō Vidyutaṃ Tatpuruṣōmānava Ssa Ēnā Nbrahmagamayatyēṣa Dēvayānaḥ Pandhā Iti ।।2।।
अथ य इमे ग्राम इष्टापूर्तेदत्तमित्युपासते ते धूम मभिसम्भवन्ति धूमा द्रात्रिं रात्रे रपरपक्षमपरपक्षाद्यान् षड्दक्षिणैति मासां स्तान्नैते संवत्सर मभि प्राप्नु वन्ति॥३॥
Atha Ya Imē Grāma Iṣṭāpūrtēdattamityupāsatē Tē Dhūma Mabhisambhavanti Dhūmā Drātriṃ Rātrē Raparapakṣamaparapakṣādyān Ṣaḍdakṣiṇaiti Māsāṃ Stānnaitē Saṃvatsara Mabhi Prāpnu Vanti ।।3।।
मासेभ्यः पितृलोकं पितृलोका दाकाशमाकाश च्चन्द्र मस मेषसोमो राजा तद्देवाना मन्नंतंदेवा भक्षयन्ति॥४॥
Māsēbhyaḥ Pitṛlōkaṃ Pitṛlōkā Dākāśamākāśa Ccandra Masa Mēṣasōmō Rājā Taddēvānā Mannantandēvā Bhakṣayanti ।।4।।
तस्मि न्यावत्सम्पात मुषित्वाथैत मेवाध्वानं पनर्निवर्तन्ते यथेत माकाशमाकाशा द्वायुं वायुर्भूत्वा धूमो भवति धूमोभूत्वाभ्रं भवति॥५॥
Tasmi Nyāvatsampāta Muṣitvāthaita Mēvādhvānaṃ Panarnivartantē Yathēta Mākāśamākāśā Dvāyuṃ Vāyurbhūtvā Dhūmō Bhavati Dhūmōbhūtvābhraṃ Bhavati ।।5।।
अभ्रं भूत्वा मेघोभवति मेघोभूत्वा प्रवर्षति त इह व्रीहि यवा ओषधिवनस्पतय

स्तिलमाषा इतिजायन्तेतोवैखलुदुर्निष्प्रपतरं योयो ह्यन्न मत्ति योरेतस्पिंचति तद्भूय एव भवति॥६॥

Abhraṃ Bhūtvā Mēghō Bhavati Mēghō Bhūtvā Pravarṣati Ta Iha Vrīhi Yavā Ōṣadhivanaspataya Stilamāṣā iti Jāyantētō Vaikhalu Durniṣprapataraṃ Yōyō Hyanna Matti Yō Rēta Spiñcati Tadbhūya Ēva Bhavati//6//

त द्य इह रमणीय चरणा अभ्याशो ह यत्ते रमणीयां योनि मापद्येरन् ब्राह्मण योनिं वा क्षत्रिययोनिं वा वैश्ययोनिं वाथ य इह कपूयचरणा अभ्याशो ह यत्ते कपूयां योनि मापद्येरन् श्वयोनिं वा सूकरयोनिं वा चण्डाल योनिं वा॥७॥

Ta Dya Iha Ramaṇīya Caraṇā Abhyāśō Ha Yattē Ramaṇīyāṃ Yōni Māpadyēran Brāhmaṇa Yōniṃ Vā Kṣatriyayōniṃ Vā Vaiśyayōniṃ Vātha Ya Iha Kapūyacaraṇā Abhyāśō Ha Yattē Kapūyāṃ Yōni Māpadyēran Śvayōniṃ Vā Sūkarayōniṃ Vā Caṇḍāla Yōniṃ Vā//7//

अथै तयोः पथोर्नकतरेण च न ता नीमानि क्षुद्राण्य सकृदावर्तीनि भूतानि भवन्ति जायस्व म्रियस्वेत्येत तृतीयं स्थानं तेनासौ लोको न संपूर्यते तस्माज्जुगुप्सेत तदेषश्लोकः॥८॥

Adhai Tayōḥ Pathōrnakatarēṇa Ca Na Tā Nīmāni Kṣudrāṇya Sakṛdāvartīni Bhūtāni Bhavanti Jāyasva Mriyasvētyēta Tṛtīyaṃ Sthānaṃ Tēnāsau Lōkō Na Sampūryatē Tasmājjugupsēta Tadēṣa-ślōkaḥ//8//

स्तेनो हिरण्यस्य सुरां पिबंश्च गुरो स्तल्प मावसन् ब्रह्महाच एते पतन्ति चत्वारः पंचमश्चाचरं स्तैरिति॥९॥

Stēnō Hiraṇyasya Surāṃ Pibaṃśca Gurō Stalpa Māvasan Brahmahāca Ētē Patanti Catvāraḥ Pañcamaścācaraṃ Stairiti//9//

अथ ह य एतानेवं पञ्चाग्नीन् वेद न सहतै रथाचरन् पाप्मना लिप्यते शुद्धः पूतः पुण्यलोको भवति य एवं वेद य एवं वेद॥१०॥

Atha Ha Ya Ētānēvaṃ Pañcāgnīn Vēda Na Sahatai Rathācaran Pāpmanā Lipyatē Śuddhaḥ Pūtaḥ Puṇyalōkō Bhavati Ya Ēvaṃvēda Yaēvaṃ Vēda//10//

य एषोक्षिणि पुरुषो दृश्यत एष अत्मेति होवाचैत दमृतमभय मेतद्ब्रह्मोति तद्यद्यप्यस्मिन् सर्पिर्वोदकं वासिंचति वर्त्मनी एव गच्छति॥

छांदोग्योपनिषद्, ६.१५.१.

Ya Ēṣōkṣiṇi Puruṣō Dṛśyata Ēṣa Atmēti Hōvācaita Damṛtamabhaya Mētadbrahmēti Todyadyapyasmin Sarpirvōdakaṃ Vāsiñcati Vartmanī Ēva Gacchati// Chāndōgyōpaniṣad, IV.15.1.

एतं संयद्वाम इत्याचक्षत एतं हि सर्वाणि नामा न्यभिसंयन्ति सर्वाण्येनं नामा न्यभिसंयन्ति य एवं वेद॥२॥

Ētaṃ Saṃyadvāma Ityācakṣata Ētaṃ Hi Sarvāṇi Nāmā Nyabhisaṃyanti Sarvāṇyēnaṃ Nāmā Nyabhisaṃyanti Ya Ēvaṃ Vēda//2//

एष उ एव नामनी रेषहि सर्वाणिनामानि नयति सर्वाणि नामानि नयति य एवं वेद॥३॥

Ēṣa U Ēva Nāmanī Rēṣahi Sarvāṇināmāni Nayati Sarvāṇi Nāmāni Nayati Ya Ēvaṃ Vēda//3//

एष उ एव भामनी रेषहि सर्वेषु लोकेषु भाति सर्वेषु लोकेषु भाति य एवं वेद॥४॥

Ēṣa U Ēva Bhāmanī Rēṣahi Sarvēṣu Lōkēṣu Bhāti Sarvēṣu Lōkēṣu Bhāti Ya Ēvaṃ Vēda//4//

अथ य दु चै वास्मि ञ्छव्यं कुर्वन्ति यदिच नार्चिष मेवाभिसम्भवन्त्यर्चिषोहरह्न आपूर्यमाण पक्षमापूर्यमाण पक्षाद्यान् षडुदज्जेति मासां स्तान्मासेभ्य स्संवत्सरं संवत्सरा दादित्य मादित्याच्चन्द्रमसं चन्द्रमसो विद्युतं तत्पुरुषोमानव स्स एना न्ब्रह्मगमयत्येष देवपथो ब्रह्मपथ एतेन प्रतिपद्यमाना इमं मानव मावर्तं नावर्तन्ते नावर्तन्ते॥५॥

Atha Ya Du Cai Vāsmi Ñchavyaṃ Kurvanti Yadica Nārciṣa Mēvābhisambhavantyarciṣōharahna Āpūryamāṇa Pakṣamāpūryamāṇa Pakṣādyān Ṣaḍudajjēti Māsāṃ Stānmāsēbhya Ssaṃvatsaraṃ Saṃvatsarā Dāditya Mādityāccandramasaṃ Candramasō Vidyutaṃ Tatpuruṣōmānava Ssa Ēnā Nbrahmagamayatyēṣa Dēvapathō Brahmapatha Ētēna Pratipadyamānā Imaṃ Mānava Māvartaṃ Nāvartantē Nāvartantē//5//

32. P.V. Kane, Vol. III, p. 923 ff; Vol. V. p. 484 ff.
33. P.V. Kane, Vol. V. p. 484 ff.
34. Manu, IX., 301-2; Mahābhārata, Śāntiparvan, 70.6.
35. Eliade, 'Cosmos and History : The Myth of the Eternal Return', p. 112, ff; 'Time and Eternity in Indian Thought', In Man and Time.
36. Viṣṇu Purāṇa, IV; Romila Thapar, 'Genealogical patterns as perceptions of the past', Studies in History, 1991, n.s., 7.1.1-36.
37. Śatapatha Brāhmaṇa, I.8.1.
38. Romila Thapar, 'Time as a Metaphor of History : Early India', Oxford India, paperbacks, p. 27.
39. Romila Thapar, 'From Lineage to State', Delhi, 1984.
40. Kane, V. 687.
41. Viṣṇu Purāṇa, IV.24. 104-7, 113.
42. Aihole Inscriptions of Pulakesin-II, Epigraphia Indica, 67; J.F.

Fleet, 'The Kaliyuga of 3102 BC', Journal of the Royal Asiatic Society, 1911, 479-96 and 675-98.

43. Viṣṇu Purāṇa, IV.24.43.
44. D.C. Sircar, 'Indian Epigraphy', Delhi, 1965, p. 251 ff; F. Kielhorn, 'Examination of Questions connected with the Vikrama Era', The Indian Antiquary, 1891, 20.397-414.
45. Romila Thapar, 'Time as a Metaphor of History : Early India', p. 34.
46. *Ibid.*
47. Kane, III, 126; D.C. Sircar, Indian Epigraphy, Delhi, 1965, p. 227.
48. Manu, III. 162; VI. 50; Baudhāyana, II, 1.2.1-6; Arthaśāstra IX.4.25-26;

षष्ट्यब्दयुगवर्षमासदिनहोराधिपतीनां प्रतिपत्तिच्छेदवित्।

बृहत्संहिता, २.६.

Ṣaṣṭyabdayugavarṣamāsadinahōrādhipatīnāṃ Pratipatticchēdavit।

Bṛhatsaṃhitā, 2.6.

ग्रन्थतश्चार्थ तश्चैतत्कृत्स्नं जानाति यो द्विजः।
अग्रभुक् स भवेच्छ्राद्धे पूजितः षङ्क्तिपावनः॥

बृहत्संहिता, २.३१.

Granthataścārtha Taścaitatkṛtsnaṃ Jānāti Yō Dvijaḥ।
Agrabhuk Sa Bhavēcchrāddhē Pūjitaḥ Ṣaṅktipāvanaḥ॥

Bṛhatsaṃhitā, 2.31.

49. Romila Thapar, 'Clan, Caste and Origin Myths in Early India', Simla, 1992.
50. Romila Thapar, 'Society and Historical Consciousness the Itihāsa Purāṇa Tradition', In S. Bhattacharya and Romila Thapar eds., 'Situating Indian History', Delhi, 1986, pp. 353-83.
51. V.S. Pathak, 'Ancient Historians of India', Bombay, 1966.
52. E.I., Vol. I, p. 122.
53. Romila Thapar, 'Time as Metaphor of History; Early India', Oxford India, paperbacks, p. 37.
54. *Ibid.*
55. Viṣṇu Purāṇa, Trans., H.S. Wilson, 22.20.4.

CHAPTER - V

PRE-KALIYUGA AND KALIYUGA ERAS

The natural divisions of Time-Years, Seasons, Months, Days have in all ages been determined by the motions of the Sun and the Moon. Sun governs the most important factor of the Time. When and how the counting of Time started, it is difficult to predict but various *Sūtras* and *Brāhmaṇas* have given valuable account in this regard. The revolution of Earth around the Sun governs calculation of years and sequence of seasons. The revolution of the Moon around the Earth governs the calculation of months and the rotation of the Earth around its axis forms the day and night. Usually, the day in Indian calendar containing astronomical and meteorological data is calculated in relation to *Tithis* and *Nakṣatras* for the religious practices.

Yuga Mentioned in as *Daśamē Yuga,* which refers to short and long periods of Time. In *Vēdas* it is referred as tenth decade. The *Brāhmaṇa* literature refers to four *Yugas,* i. e., *Kṛta, Trēta, Dvāpara* and *Kali*. The terms *Saṃvatsara, Parivatsara, Iḍā-Vatsara, Id-Vatsara* and *Vatsara* are referred in *Yajur Vēda*. The number of years in the *Yugas* vary according to different versions as shown:

According to	Kṛta Yuga	Trēta Yuga	Dvāpara Yuga	Kali Yuga
Brahmagupta	1,728,000	1,296,000	864, 000	432,000
Āryabhaṭṭa	1,080,000	1,080,000	1,080,000	1,080,000
Purāṇas	1,440,000	1,080,000	720,000	36,000

The most ancient year probably consisted of 360 days which approximated roughly to twelve revolutions of the Moon and one of the Sun. In one of the hymns of the *Yajurveda,* the Sun's annual course through the heavens is described as his twelve spoked wheel.[1] The 360 days with as many nights are called his 720 children. In another part of the same hymn the Sun's annual course is somewhat differently described: "The felloes are twelve, the wheel is one, three are the axles: within it are collected 360 spokes".[2] Here the spokes represent the number of days; the axles are three seasons of Heat, Rain and Cold; and the twelve felloes are the twelve months.

As stated earlier, the Years are reckoned by the course of the Sun, while the Months are regulated by the revolutions of the Moon and to that extent Solar and Lunar calendars have come into existence. The Solar year commences with the instant of the Sun's entrance *(Saṅkrānti)* into the Hindu sign of Mēṣa-Aries, which is at the same time the beginning of the Solar month *Vaiśākha.* The beginnings of the other Solar months are similarly determined by the entrance of the Sun into the different Zodiacal signs viz., Vaiśākha, *Jyēṣṭha, Āṣāḍha, Śrāvaṇa, Bhādrapada, Aśvina, Kārtīka, Mārgaśira, Pauṣya, Māgha, Phālguṇa* and *Caitra.*

Further, there are *Āyanas* known as *Uttarāyaṇa* and *Dakṣiṇāyana* basing upon the period of the Sun's progress North or South of the ecliptic. Six months form an *Āyana* and two *Āyanas* compose a Year.

The Year in India is generally divided into six seasons: Spring *(Vasanta)*, Summer *(Grīṣma;* Hot season), the Rainy season *(Varṣa;* Monsoon), Autumn *(Śarad),* Winter *(Hēmanta)* and the Cold season *(Śiśira,* which comes between Winter and Spring). Each season consists of two Months, and three *Ṛtus* compose one *Ayana* or half of Solar year.

Ṛgvēda refers to *Māsa* (month) quite repeatedly. Since the orbit of the Earth ecliptic, the duration of Solar months varies from 29 to 30 days. In case when two Lunar months begin in a Solar month they are denoted as *Adhika* and *Nija.* These Sanskrit Zodiac signs are frequently mentioned as *Mēṣa, Vṛṣabha, Mithuna, Karkāṭaka, Simha, Kanya, Tula, Vṛścika, Dhanus, Makara, Kumbha* and *Mīna.* Months are of two types, Solar and Lunar. Solar month is the time required for the Sun to pass from one sign to another sign. The Lunar month is the time taken for one full Moon to the next. Thus a Lunar month corresponds to one Lunation. It is reckoned either from New Moon to New Moon, or Full Moon to full Moon. The first scheme is called the *Amānta, Darśānta* or Southern scheme; the latter, the *Pūrṇimānta* or Northern scheme.

Each Month consists of two *Pakṣas,* i. e., Fortnight which is based on the revolution of the Moon. The Bright Fortnight *(Śukla, Śuddha* or *Sita Pakṣa* or *Suḍi, Sudi, Suti)* is the period of the waxing Moon; the Dark Fortnight *(Kṛṣṇa, Bahuḷa* or *Asita Pakṣa,* or *Baḍi, Vaḍi, Vaṭi)* that of the waning Moon. As indicated above the Bright Fortnight in the *Amānta* or Southern scheme is the first *Pakṣa* of the Moon; in the *Pūrṇimānta* or Northern scheme, it is the last. But in either case it denotes the same space of time. It is different with the Dark Fortnight; for the Dark Fortnight of an *Amānta* month corresponds to that of the following month in the *Pūrṇimānta* scheme, e. g., the Dark Fortnight of *Caitra* in the *Amānta* scheme is the Dark Fortnight of *Vaiśākha* in the *Pūrṇimānta* scheme and vice versa.

The name of the Lunar month is now invariably determined by New Moon forming the true beginning of its Bright Fortnight. For the Lunar month takes the name of the Solar month in which that New Moon occurs, e.g., the New

Moon in the Solar Month of *Caitra* always inaugurates the Bright Fortnight of the Lunar month *Caitra*. If two Moons occur within one Solar month there will be two Lunar months of the same name: The proper one *(Nija)* and the intercalated one *(Adhika)*. In the South, the intercalated month precedes the proper one; in the North, it is inserted between the two *Pakṣas* of the proper month. Usually, however, the two homonymous *Pakṣas* are marked *Prathama* and *Dvitīya*. If no New Moon occurs in a Solar month, there will of course be no Lunar month of that name, and that month is considered expunged *(Kṣaya)*.

Each *Pakṣa* is divided into fifteen *Tithis*. A Tithi is the time required by the Moon to increase its distance Westward from the Sun by twelve degrees of the Zodiac. The *Tithis* are named or numbered by Sanskrit ordinals *Prathama, Dvitīya,* up to *Pañcadaśi,* but the fifteenth *Tithi* of the bright half is also called the Full Moon Tithi-*Pūrṇamasi,* and the fifteenth *Tithi* of the dark half, the New Moon Tithi-*Amavasi* or *Amāvāsya;* and the first *Tithi* of either half bears the name *Pratipad* or *Pratipada*. The instants of New and Full Moon are the terminal points of the Dark and the Bright Fortnights. In civil reckoning, the Tithis are coupled with the Civil days in such a way that the Civil day (from true Sunrise to Sunrise) takes the name, i. e., number of that *Tithi* which ends in it.

In connection with the civil reckoning it may be stated that Hindus have adopted the planetary Week and the Week days are named in the order as *Ravivāra, Sōmavāra, Maṅgaḷa* or *Bhaumavāra, Budhavāra, Guruvāra* or *Bṛhaspativāra, Śukravāra* and *Śanivāra*. The *Vāra* is also sometimes called as *Dina* or *Divasa* or Solar day is counted in regular order from *Ravivāra* as 1 up to *Śanivāra* as 7 or zero. The mean Civil day is divided into 60 *Ghaṭikas* of 60 *Palas* each. The *Ghaṭika* is the 60th division of the 24 hours of a Solar day and night, which is always reckoned from Sunrise to Sunrise and is therefore equal to 24 minutes. A *Pala* is the sixtieth division of a *Ghaṭi* is therefore equal to 24 seconds.

Further, the Hindu calendar contains the *Nakṣatra* or Constellation beginning from *Kṛttika* to *Bharaṇi;* the *Yōga* or propitious or unfavourable time for various activities (determined by the relation of the Sun and the Moon), with the *Karaṇas* (since each *Tithi* has two *Karaṇas,* there are 60 *Karaṇas* in the Lunar month); then the *Rāśi* or Signs of the Zodiac.

ANCIENT ERAS PRIOR TO KALIYUGA

Saptaṛsi Kāla (or) Cycle of the Seven-Ṛṣis

The oldest Eras described by the astronomers are the *Saptaṛṣi-Kāla,* or the Cycle of the *Seven Ṛṣis;* the *Bārhaspatya-mānas,* or sixty and twelve year Cycles of Jupiter; and the *Kaliyuga,* or beginning of the Kali-age. The oldest of them, the *Saptaṛṣi-Kāl* ascends only to B.C. 4077, or perhaps to 6777 B.C., while the *Bārhaspatyamāna* and the *Kaliyuga* which only a little beyond 3000 B.C. The *Saptarṣi-Kāl* is unknown in Southern India, but the *Kaliyuga* and the sixty-year Cycle of Jupiter are well known. The eras of *Buddha* and *Mahāvīra,* both of which are prior to *Vikramāditya,* must have been used by the Buddhists and Jains at an earlier period. The former was certainly current among the Buddhists in the time of *Aśōka* and the latter was probably in use about the same period. In the Mathura Inscriptions of the Indo-Scythian kings which are found upon the statues of both Jains and Buddhists the dates are invariably expressed in an Era which may have originated with *Kaniṣka,* but which was most probably an Indian adoption of the Seleucedian Era as suggested by Mr. Thomas. In dealing with Indian-Greeks, there is one fact that every year that is mentioned by number refers to years actually elapsed.

The *Sapt-Ṛṣi-Kāla* or 'Cycle of the *Seven Ṛṣis*' called also *Saptarṣi* and *Sat-Rikhi-Kāla* is so named after the seven stars of the Constellation of the Great Bear. It is the only mode of reckoning employed in the *Rājataraṅgiṇī* or History of Kāśmīr, and it is still used in the hill states to the South-East of Kāśmīr between the *Cīnāb* (Chenab) on the West and the *Jamuna* on the East.

The *Sapt-Ṛṣi-Kāl* or the Era of the *Seven Ṛṣis* is also called as *Lōka-Kāla,* or Common Era, is a cycle of 2700 years divided into 27 centenary periods, a new reckoning being started at the beginning of each Century. The theory of the cycle is that the *Seven Ṛṣis* or Stars of *Ursa Major,* remain for one century in each of the 27 *Nakṣatras* or Lunar mansions. All authorities agree in making *Aśvini* the first of the *Nakṣatras* and in stating that the *Mahābhārata* took place when the *Ṛṣis* were in the Lunar Constellation *Magha,* the tenth of the series. The *Purāṇas,* and the practice of all the people who still use this cycle, excepting only the *Kāśmīris* agree in making the Era of *Yudhiṣṭhira* as the same as the *Kaliyuga.* All however, agree in stating that at the time of *Mahābhārata,* the Seven Ṛṣis had already passed seventy-five years in *Magha*. But as *Varāha* places the great war 653 years after the beginning of *Kaliyuga,* or in 2449 B.C., that year should have been the *76th* of the *10th Nakṣatra* and the *976th* year of the cycle. This would fix the *1st* year of the centenary period to the *25th* year of each century B.C. and to the *76th* year of each century A.D. But to prevent the confusion that would have thus arisen, *Varāha* simply ignores the generally accepted belief that the *Ṛṣis* had spent 75 years in *Magha* when the *Mahābhārata* took place and retained from 3177 (or 3102+75) to B.C. 2477. Accordingly, Varāha's followers place the initial point of the *Vṛhaspati Cakra* in 3377 B.C. in Aśvini, so that each century begins in the *26th* year of each century of the *Kaliyuga* exactly as Dr. Bahler has informed. This also accords with *Kāśmīri* sources that the *Ṛṣis* had completed three revolutions less 25 years in the *Dvāpara Yuga* before the *Kali Yuga* began, i.e., the Cakra preceded the *Kaliyuga* by 275 years equivalent to B.C. 3377 or 3102+275 years. The following Table[3] on Cycle of Seven Ṛṣis shows the initial year of each Century, as well as its *Nakṣatra* or Lunar asterism according to the different reckoning of *Vṛddha Garga* and the *Purāṇas* on one hand, and of *Varāha* and the later astronomers on the other. (For Table of Cycle of the *Seven Ṛṣis*-Refer next Page).

The numbers placed in the table against the asterisms show the number of each Century, while the beginning of the Cakra or complete cycle of 2700 is indicated by the number 1 placed against *Aśvini*. Thus on the left hand it will be seen that the cycle of the commonly received account began in the years 6777, 4077 and 1377 B.C., and in 1325 A.D., while those of *Varāha Mihira's* reckoning began in 3377 and 677 B.C. By the former it will be seen that the *Seven Ṛṣis* were in *Magha* between 3177 and 3077 B.C.

Bārhaspatyamāna—60-Year Cycle of Jupiter

Bārhaspatyamāna or Cycle of Jupiter is a period of 60 years of five revolutions of the planet, each year of which has a different name. There are three different modes of reckoning the Cycle of sixty years, of which the oldest is certainly that preserved by *Varāha Mihira* as the first year of the *Kaliyuga* by his account is the twenty-seventh year of the Juvian Cycle. The second is the reckoning of *Jyōtistatva,* which is clearly only a correction of *Varāhamihira's* method as it makes the first year of the cycle, corresponds with the first year of the

Cycle of the Seven Ṛsis

According to Vṛddha Garga and the Purāṇas		Lōka-Kāla or Saptaṛṣi Cakra Initial years of Centuries				According to Varāha Mihira and the later Astronomers	
		B.C.	B.C.	B.C.	A.D.		
1.	Aśvini	6777	4077	1377	1325	U. Aṣāḍha	21
2.	Bharaṇi	6677	3977	1277	1425	Śravaṇa	22
3.	Kṛttika	6577	3877	1177	1525	Dhaniṣṭha	23
4.	Rōhiṇi	6477	3777	1077	1625	Śatabhiṣa	24
5.	Mṛgaśira	6377	3677	977	1725	P. Bhādra	25
6.	Ārdra	6277	3577	877	1825	U. Bhādra	26
7.	Punarvasu	6177	3477	777	1925	Rēvati	27
8.	Puṣya	6077	3377	677	2025	Aśvini	1
9.	Āślēṣa	5977	3277	577	2125	Bharaṇi	2
10.	Magha	5877	3177	477	2225	Kṛttika	3
11.	P. Phalguṇi	5777	3077	377	2325	Rōhiṇi	4
12.	U. Phalguṇi	5677	2977	277	2425	Mṛgaśira	5
13.	Hasta	5577	2877	177	2525	Ārdra	6
14.	Citra	5477	2777	BC. 77	2625	Punarvasu	7
15.	Svāti	5377	2677	AD. 25	2725	Puṣya	8
16.	Viśākha	5277	2577	125	2825	Āślēṣa	9
17.	Anūrādha	5177	2477	225	2925	Magha	10
18.	Jyēṣṭha	5077	2377	325	3025	P. Phalguṇi	11
19.	Mūla	4977	2277	425	3125	U. Phalguṇi	12
20.	P. Āṣāḍha	4877	2177	525	3225	Hasta	13
21.	U. Āṣāḍha	4777	2077	625	3325	Citra	14
22.	Śravaṇa	4677	1977	725	3425	Svāti	15
23.	Dhaniṣṭha	4577	1877	825	3525	Viśākha	16
24.	Śatabhiṣa	4477	1777	925	3625	Anūrādha	17
25.	P. Bhādra	4377	1677	1025	3725	Jyēṣṭha	18
26.	U. Bhādra	4277	1577	1125	3825	Mūla	19
27.	Rēvati	4177	1477	1225	3925	Pūrvāṣāḍha	20

Kaliyuga. Both of these reckonings have been in use in Northern India where the necessary omission of the every sixth year of the Juvian Cycle has always been preserved.

There is difference of reckoning of 60-year Cycle of Jupiter in *Jyōtistatva* and *Sūrya Siddhānta*. In fact the difference between the two is never more than one year and that only between the two periods of omitted years. In the second century A.D., the omitted year of the *Sūrya Siddhānta* reckoning took place 136, while that of the *Jyōtistatva* was two years later in 138. In A.D. 394, the omissions took place together. In A.D. 479, the *Jyōtistatva* omitted year preceded that of the *Sūrya Siddhānta* by one year, but in the present century the *Jyōtistatva* omitted year, 48 in 1848, preceded tne other No. 1 of 1856, by 13 years. The current years of the two cycles, however, generally correspond, excepting in the short periods between the two omissions, when they differ by only one year. The years 847 and 907 A.D. were initial years of cycles in all three modes of reckoning and the numbers of all the years coincided from A.D. 825 (the 39th year) down to A.D. 909.

In the table[4] given the Sanskrit names of all the sixty years are given with their numbers counting from *Prabhava*. And also the translations of the Tibetan names, which were derived from, the Chinese are given for all the sixty years. They are formed by a combination of the names of the twelve animals of the smaller cycle of twelve years with the five elements.

Bārhaspatyamāna—12 Year Cycle of Jupiter

The smaller Cycle of Jupiter consists of a period of 12 years or one fifth of the Greater Cycle. *Varāhamihira* notices within the following term : "Each year (during which Jupiter completes a twelfth part of his revolution) has to bear the

60 Year Cycle of Jupiter—Bārhaspatya Cakra
Names of the 60 Years of the Jovian Cycle

No	Sanskrit	Tibetan	No
1.	Prabhava	Fire-Hare	4
2.	Vibhava	Earth-Dragon	5
3.	Śukla	Earth-Serpent	6
4.	Pramōda	Iron-Horse	7
5.	Prajāpati	Iron-Sheep	8
6.	Aṃgīrasa	Water-Ape	9
7.	Śrīmukha	Water-Bird	10
8.	Bhava	Wood-Dog	11
9.	Yuvan	Wood-Hog	12
10.	Dhātar	Fire-Mouse	13
11.	Īśvara	Fire-Ox	14
12.	Bahūdhānya	Earth-Tiger	15
13.	Pramāthin	Earth-Hare	16
14.	Vikrama	Iron-Dragon	17
15.	Vṛṣa	Iron-Serpent	18
16.	Citrabhānu	Water-Horse	19
17.	Subhānu	Water-Sheep	20
18.	Tāraṇa	Wood-Ape	21
19.	Pārthiva	Wood-Bird	22
20.	Vyaya	Fire-Dog	23
21.	Sarvajit	Fire-Hog	24
22.	Sarvadhārin	Earth-Mouse	25
23.	Virōdhin	Earth-Ox	26
24.	Vikṛta	Iron-Tiger	27
25.	Khara	Iron-Tiger	28
26.	Nandana	Water-Dragon	29
27.	Vijaya	Water-Serpent	30
28.	Jaya	Wood-Horse	31
29.	Manmadha	Wood-Sheep	32

No	Sanskrit	Tibetan	No
30.	Durmukhi	Wood-Ape	33
31.	Hēvaḷamba	Fire-Bird	34
32.	Vilambin	Earth-Dog	35
33.	Vikārin	Earth-Hog	36
34.	Śārvari	Iron-Mouse	37
35.	Plava	Iron-Ox	38
36.	Śōbhakṛt	Water-Tiger	39
37.	Śubhakṛt	Water-Hare	40
38.	Krōdhin	Water-Dragon	41
39.	Viśvāvasu	Wood-Serpent	42
40.	Parābhava	Fire-Horse	43
41.	Plavaṅga	Fire-Sheep	44
42.	Kīlaka	Earth-Ape	45
43.	Saumya	Earth-Bird	46
44.	Sādhāraṇa	Iron-Dog	47
45.	Rādhākṛt	Iron-Hog	48
46.	Parīdhāvi	Water-Mouse	49
47.	Pramādin	Water-Ox	50
48.	Ānanda	Wood-Hare	51
49.	Rākṣasa	Wood-Tiger	52
50.	Anala	Fire-Dragon	53
51.	Piṅgaḷa	Fire-Serpent	54
52.	Kālayukta	Earth-Serpent	55
53.	Siddhārtha	Earth-Horse	56
54.	Raudra	Iron-Ape	57
55.	Durmati	Iron-Bird	58
56.	Dundubhi	Water-Dog	59
57.	Udgārin	Water-Dog	60
58.	Raktākṣa	Wood-Mouse	1
59.	Krōdhana	Wood-Ox	2
60.	Akṣaya	Wood-Tiger	3

name of the Lunar mansion in which he rises. The years follow each other in the same order as the Lunar months". They are also named after the Lunar months with the prefix of the word *'Mahā'.* Thus, Lalla says : *'Maghā Ca Māghāyāṃ Yukta Māghāyāṃ Ca Gururgaḍa Mahā Māgha'*. "When both the Moon and Jupiter are in the asterism *Maghā* on the day of Full Moon of the month *Māgha,* then the year is called *Mahā Māgha*".

The statement of *Varāha* quoted above that the year has to bear the name of the mansion in which Jupiter rises can be further explained like this. The twenty seven *Nakṣatras* or Lunar mansions are divided into twelve groups, nine of which comprise two mansions only and the remaining three, each three mansions. One *Nakṣatra* in each of these twelve groups gives its name to the Luni-Solar months and consequently to the years of this cycle.

According to the rule for naming the several years of the twelve years of the Jupiter, the year is called after the *Nakṣatra* in which the planet rises heliacally. But in practice, the names of the Jovian years are met to coincide with those of the Luni-Solar months. So that should the planet rise in *Bharaṇi* the year is not called *Bharaṇi,* but *Aśvini* which is the name giving *Nakṣatra* of the group to which *Bharaṇi* belongs.

Garga mentions that 172 of Jupiter years were equal to 170 Solar years, and the two names of *Āśvayuja* and *Caitra* must be omitted. This proportion was afterwards, altered by *Varāha* who made 172 years of Jupiter equal to 170 Solar years on which account two of *Bṛhaspati's* years are to be omitted in that period. So, practically every 86th name is expunged and consequently the omissions are confined to six names out of twelve or in other words, the omissions fall only on the alternate names in regular succession. Thus the six omitted names are *Śrāvaṇa Āśvayuja, Mārgaśira, Māgha,*

Caitra and *Jyeṣṭha*. According to Cunningham '...the two cycles years were subject to omission because *Caitra* and *Āśvayuja* alone are mentioned. My impression is that the same six months that are omitted by *Varāha's* rule were also subject to omission in *Garga's* time. But even admitting that *Caitra* and *Āśvayuja* were the only two years in *Varāha's* scheme, they will of course, remain constant, as the average period of omission is in both cases the 86th year. If then we accept the year 310 A.D. in which *Caitra* was omitted as common to both systems; we have only to take the *Āśvayuja* and *Caitras* which fall nearest to the 85-year periods, either those preceding (A) or those following them (B) and the result will be the same excepting only as regards the names of the other omitted years.[5] Further, he gives this arrangement of names in the following table.

Varāhamihira			Proposed Arrangements					
A.D.	Interval		A.D.	Interval	A	A.D.	Interval	B
310		Caitra	316		Caitra	310		Caitra
	85			77			89	
395		Jyēṣṭha	387		Āśvayuja	399		Āśvayuja
	85			89			89	
480		Śrāvaṇa	476		Caitra	488		Caitra
	85			89			77	
565		Āśvayuja	565		Āśvayuja	565		Āśvayuja
	85			77			89	
650		Agrahāyaṇa	642		Caitra	654		Caitra
	85			89			89	
735		Māgha	731		Āśvayuja	743		Āśvayuja
	85			89			77	
820		Caitra	820		Caitra			Caitra
/6	510	Years	/6	510	Years	/6	510	Years
Mean	85	Interval	Mean	85	Interval	Mean	85	Interval

From the above table it will be seen that a regular succession of *Caitras* and *Āśvayujas* might be omitted while still retaining a uniform mean period of 85 years. It will also be seen that at every third period, the names of the omitted years, as well as the dates of omission, agree with those of *Varāhamihira*.

Yudhiṣṭhira and Bhārata Yuddha Eras

The usual eras, by which our Hindus assign dates to events are the *Vikramārka* and *Śālivāhana* otherwise known as *Saṃvat* and *Śaka* respectively. Of these, the latter begins 78 years after and the former 56 years before the Christian Era. They did not adopt the system of retrograde calculation but used other eras, which began many thousand years before it. Two of these are the *Yudhiṣṭhira* Era and *Kaliyuga*. Besides these, the same writers reckoned their date according to yet another era, that is *Bhāratayuddha Era*. One understands the importance of these eras to the history of India, when one comes to know that the *Kāśmīr* chronology of the *Rājataraṅgiṇi,* the *Magadha* chronology, as well as the chronicles of the Solar and Lunar races of the *Purāṇas*—these and other chapters of the ancient history of India chiefly base themselves on the above named ancient eras.

At present, however, our modern astrologers and others make use only of the *Kaliyuga* along with *Vikramārka* and *Śālivāhana Eras*. But they have certain data from which they can calculate the other two ancient eras, viz., *Yudhiṣṭhira* and *Bhāratayuddha Kāla* and the data are such the results of their calculation show that these eras are identical with *Kaliyuga* in respect of their beginning. According to the astrologers, *Yudhiṣṭhira Era* is obtained by adding 3049 to the *Vikrama Saṃvat,* which in its turn is got, by adding 135 to the *Śaka* date. According to *Pañcāṅgasārani* the date of the *Yudhiṣṭhira Era* is also obtained by adding 3129 to the

Śaka Era, i.e., the Śaka begins with the 3180th year of the Yudhiṣṭhira Era. Again on a hill near *Aihōle* (Bījāpūr district), there is an inscription in a *Jaina* temple of *Pulakēśin-II* of the *Cāḷukya* family in which it is stated that the temple was erected 3735 years after the *Mahābhārata* battle and when 556 years of the *Śaka Era* had passed. Hence, it is clearly seen that the great battle is supposed to precede the *Śaka Era* while 3179 years (3735-556), that is the *Śaka* begins with the 3180th year of the *Bhāratayuddha Era* also.

Now, *Bhāratayuddha Era,* as the name itself explains must naturally be reckoned from the date of the Great Battle, while the *Yudhiṣṭhira Era* must evidently be reckoned from *Yudhiṣṭhira's* ascending the throne of the kingdom. *Vāyu, Viṣṇu, Matsya* and other *Purāṇas* are unanimous in declaring that *Kaḷiyuga* begins on the very day of *Kṛṣṇa's* decease. Therefore, it is obvious that the *Yudhiṣṭhira Era* must have been older than *Kaliyuga* by 35 years; and that the Great Battle must be assigned a date 15 years before the *Yudhiṣṭhira Era,* or in other words half-a-century before *Kaliyuga.*

Eras of Kali-Yuga

The *Kaliyuga,* or fourth age of Hindu chronology dates from the year 3102 B.C.; the year 1 expired or completed, being B.C. 3101. The Era of *Kaliyuga* was in use down to the time of *Varāhamihira,* who first introduced the use of the Śaka era into astronomical works. *Āryabhaṭṭa,* who was not more than fifty years prior to him, still completed by the era of the *Kali-Yuga.* The initial point of the era seems to have been a traditional date of the period of the Great War, which had been handled down perhaps for ages. This date of 3102 B.C., as the year 0 of the *Kaliyuga* was accepted by all; and from it the calculations of *Āryabhaṭṭa* and *Varāhamihira* for the Solar and Luni-Solar periods were computed.

Where the *Kaliyuga Era* is used alone, the day of the month may be expressed either according to the Solar calendar, or to the Luni-Solar one. Frequently the year is given in two different eras; one of which may be usually connected with Solar calendar and the other with the Lunar. In the North of India, the *Kaliyuga* and the *Śaka* years are generally, but not always, connected with the Solar reckoning, while in the South of India, the *Śaka Era* is usually accompanied with the Luni-Solar reckoning. The *Saṃvat* of *Vikramāditya* is the only era that is exclusively Luni-Solar.

Death of Mahāvīra

The Jains make use of an era dating from the death of their last teacher *Mahāvīra*. According to the *Śvētāmbara* sect, this event took place 470 years before *Vikrama* or in B.C. 527. The same date is used throughout the *Thērāvaḷi* of *Mērutuṇga* who says : 'Before commencement of the reign of *Vikrama, Śrī Vīra's* death took place 470 years.'[6]

Nirvāṇa of Buddha

The *Nirvāṇa,* death of *Buddha* is one of the important events in Ancient Indian History. The *Nirvāṇa* of *Buddha* has been in use from a very earlier date down to the present day. According to the Buddhist chronicles of Ceylon and Burma, the *Nirvāṇa* took place in 544 B.C. But, as the inauguration of *Aśōka* is referred to the year 218 after the *Nirvāṇa,* it seems probable that there must be an error in the date of the *Nirvāṇa* itself to the extent of 266 years as the chronology of the reign of *Aśōka* is now pretty well ascertained. His father's death took place in the year 214 of the *Nirvāṇa,* or B.C. 264, and his inauguration as king four years later after he had prevailed over his brothers.

Only two inscriptions have been found which are dated in this era. The first is contained in the Rock Edicts of *Aśōka* at *Rūpanāth* and *Sahsarāṃ*. The second occurs in an inscribed slab, which was found in the temple of *Sūrya* in the city of *Gaya*. The date of *Aśōka's* inscription is the year 256, or the 42nd year after the death of his father; his own reign being stated in the chronicles at 4 years plus 37 years, or altogether 41 years complete, and 42 current. The second date is 1813 of the *Bhāgavat Parinirvṛtte Saṃvat*, or *Nirvāṇa*, or Thursday the 1st of *Kārtīka-Bādi*.

A similar antiquity was still claimed as late as the twelfth Century A.D., during the reign of *Aśōka Balla Dēva*. Two of his inscriptions are dated in the years 51 and 74 of the *Lakṣmaṇa Sēna Era*, or in A.D. 1157 and 1180. A third inscription which is dated in the year 1813 of the *Parinirvṛtte* of *Bhāgavata*, shows that the time of the *Nirvāṇa* was believed to have occurred was about 656 to 633 B.C. However, according to the Buddhist Chronology of Ceylon and Burma the year 544 B.C. is accepted as the date of *Nirvāṇa* of *Buddha*.

Era of the Seleukidae

The intial point of the *Seleukidan Era* has been fixed[7], by Fynes Clinton to the 1st of October 312 B.C. in the beginning of Olympiad XVII, 1. This era dates from the defeat of Niknor, the General of Antigonus, by Seleukus, who thus became the Master Babylon in Olympiad XVII, 1. The Syro-Macidonian Calendar of the Seleukidae was Luni-Solar and was used in the North-West of India during the period of Indo-Scythian rule.

Era of Pārthia

The notice of a *Parthian Era* was discovered[8] by George Smith amongst the Cuneiform record at Babylon.

Vikrama Era

The well-known historical Eras of India are the *Vikrama* and the *Śaka Eras*. The *Vikrama Era* is also called as *Vikramāditya Saṃvat* is reckoned from the Vernal Equinox of the year 57 B.C. and the completion of the *Kaliyuga* year 3044. It is used all over Northern India except in Bengal, where the *Śaka Era* has been generally adopted. It is used also in *Telaṅgāṇa* and *Gujarāt*. But in latter province the year does not begin until 7 months later than in the North or with the first of *Kārtīka-Śudi* which now falls during October, but which at the beginning of the *Christian Era,* fell between the middle of September and the middle of October. The era is said to have been established by *Vikramāditya* a king of Ujjain to commemorate his victory over the *Śakas*. At present the *Vikramāditya Era* begins with the first of *Kārtīka*.

Śaka Era

The *Śaka-Kāla,* called also *Śaka-Bhūpa-Kāla* and *Śakēndra-Kāla* or the 'Era of the *Śaka* king', is perhaps more widely used than any other era. The name of the *Śaka* king was *Śālivāhana* and accordingly the era is now very generally called, *Śaka-Śālivāhana*. It is also known as *Śaka-Saṃvat*. The reckoning of the *Śaka Era* begins with the Vernal Equinox of the *Kali-Yuga* year 3179, or A.D. 78. But as the Indian count only by completed years, the year 1 begins with the Vernal Equinox of *Kaliyuga* 3180, or A.D. 79. In Northern and Southern India it is usually employed along with the Luni-Solar calendar. In converting *Śaka* dates into Christian reckoning, 78 years must be added to the given date; and vice versa to convert Christian into *Śaka* reckoning, 78 years must be deducted from the former.

Gupta Era

The *Gupta-Kāla* or *Gupta Era* is not mentioned by any native writer although it is found in several ancient inscriptions as well as on the coins of the Gupta kings. It is however noticed by *Abū Rihān* who dates it from the epoch of their extermination and of confounding it with the Era of *Ballabhi*. Now the intial point of the *Vallabhi Era* is known absolutely from Colonel Tod's inscription which makes the year 1=319 A.D., which is precisely the same date that is assigned to it by *Abū Rihān* who says that it is posterior to *Śaka* by 241 years, or 241+78=319 A.D. So it is clear that Guptas must have reigned before A.D. 319. The conclusion about the two eras has probably arisen from the fact that the *Ballabhi* kings in all their Copper plate grants continued to use the Gupta Era instead of making use of the *Vallabhi* Era itself.

Cēdi/Kālacuri Era/Kālacuri Saṃvat

There is considerable number of inscriptions of the *Kālacuri Rājas* of *Cēdi* with various dates but the actual name of the era was discovered by Mr. Begler in several inscriptions from the district of *Rāypūr* to the East of *Nāgapūr*. In some it is named *Cēdi-Saṃvat* and in others the *Kālacuri-Saṃvat*. According to Cunningham, the initial point of the era must have been A.D. 249.

Vallabhi Era

The initial point of the Era of *Vallabhi* or *Vallabhi-Kāla* is fixed by the account of *Abu-Rihān* as well as by the other dates recorded in Tod's Inscriptions to the year 319 A.D. According to the former, it was 241 years posterior to the *Sāka* or 78+241=319 A.D. According to the Tod's Inscription, Sunday the 13th *Āṣāḍha-Bādi* of the year 945 of *Śrīmad Ballabhi,* fell in the year 662 of Muhammed, 1320 of *Vikrama,* and 151 of the *Śiva Singha Saṃvat.*[9] The first year of the *Ballabhi Era* was fixed by Tod by deducting 975 from 1320, which give 375 of the *Vikrama Saṃvat* as the year 1 of the *Ballabhi Saṃvat*. Then deducting 56 from 375, he obtained 319 A.D. as the equivalent in the Christian Era. According to H.G. *Śāstri, Vallabhi Era* means the *Gupta Era* with the modified system of *Kārtīkādi* years adopted by the *Vallabhi* kingdom of *Sourāṣṭra*.[10]

Gānga Era

The question of the epoch of the *Gāṅga Era* has been discussed by several scholars and various dates ranging from A.D. 349-52 to A.D. 877-78 have been proposed as marking the foundation of the era. The Paleography of early *Gāṅga* grants dated in that era shows that it must have been founded sometime about the end of the fifth and beginning of the sixth Century AD. However, according to Prof. *Mirāṣi,* the *Gāṅga Era* began in the expired *Śaka* year 420 (A.D. 498-99), probably on the first *Tithi* of the Bright Half of the *Amānta Caitra*.[11] To convert the date *Gāṅga Era* into the *Śaka Era* we have to add 419 if the *Gāṅga* year was current, and 420 if it was expired.

Śrī Harṣa Era

Śrī Harṣa-Kāla or Era of *Śrī Harṣa* was established by the famous King *Śrī Harṣavardhana* from the first year of his reign. It was used in *Mathura* and *Kanauz*. And *Abū Rihān* is the only one who mentions this era and gives its initial point from the almanacs of *Kāśmīr* as 664 years posterior to *Vikramāditya,* or 664-57=607 A.D.

Regnal Years

Apart from the system of eras, the use of Regnal Years has been adopted by the kings of various dynasties and individual rulers and also by the members of the ruling class and by the public in general, so while issuing grants and gifts as well as statements of events pertaining to politics and administration. It is pertinent to note that the usage of dating in Regnal Years along with the eras was adopted as observed in various inscriptions by the rulers of various dynasties. It is also observed from some of the inscriptions that the subjects of a king used the Regnal Years of their king.

The dating in inscription is precise and includes the Regnal Year of the ruler, the Season, Month, *Fortnight* and the Day.

REFERENCES

1. Alexander Cunningham, 'Book of Indian Eras' Calcutta, Thackar, Spink and Co., 1883.
2. Ibid.
3. Ibid., p. 16.
4. Ibid., p. 25.
5. Ibid., p. 29-30.
6. Dr. Bhandaji, Bombay Asiatic Society Journal, IX, p. 119.
7. Alexander Cunningham, 'Book of Indian Eras', p. 38.
8. Ibid., p. 46.
9. Ibid., p. 63.
10. Śāstri, Dr. Hari Prasād, G. Article 'The Vallabhi Era', published in I.H.Q. Vol. XXIV, ed., by N.N.Law, Caxton Publications, Delhi, 1985, p. 242.
11. Indian Epigraphy, Vol. XXVI, p. 330.

CHAPTER - VI

SIGNIFICANCE OF APPLICATIONS OF TIME

According to the writers of *Dharmaśāstra*, the privileges, obligations and responsibilities of each person from the king to the common man have been fixed elaborately and we find there from that *Dharma* as observed by the authors of *Dhamaśāstra* is not a creed or religion but a mode of life and *Dāna* is an ingredient of *Dharma*.

Dāna is a gift which consists in the cessation of one's ownership over a thing and creating the ownership of another over that thing and this last course has been when the other accepts the thing. According to the works of *Dharmaśāstra* there are six constituent elements of *Dāna*. They are, the Donor, the Donee, the Charitable Attitude, the Subject of the Gift, a Proper Time and a Proper Place. An important word that deserves consideration in the subject of *Dāna* is Proper Time. A Gift *(Dāna)* must be made on the first day of each *Āyana* (the Sun's passage to the North or the South), at the beginning of *Ṣaḍaśīt* and while on eclipse of the Sun or Moon is in progress and that the rewards of gifts made on these occasions are inexhaustible. *Mahābhārata* says, a gift made on the *Amāvāsya* imparts rewards a hundred times of the reward of making it on any ordinary day, a thousand times when made on the suppression of a *Tithi,* a hundred thousand times when made on a equinoctial day *(Viṣuvat Kāla)* and a gift brings endless rewards when made on *Vyatīpāta*.

The above rules and the special occasions for gifts are amply borne in epigraphical records. When a king makes a gift of land or bestows a *Nibandha,* (1) he should execute a writing for the information of the future good king. (2) he should issue a permanent edict bearing his signature on a Copper Plate or on a stone marked at the top with his seal, (3) write down there in, the names of the ancestors and of himself, (4) the extent of the land or measurement of what is gifted, (5) and set out the passages that condemn the resumption of gifts, (6) the edict should bear the names or signatures of the royal officers such as *Agna Dautaka,* (7) that it should state to last till the Sun and Moon endure, (8) that it was not to be resumed or taken back, (9) that was to be free from all taxes, (10) that it was to go from generation to generation, (11) that it should bear king's seal, the year, the month, the fortnight and the day on which the gift was made, (12) it should be addressed to the king's officers and to all householders, etc. All these details can be seen in the tabular statement given.

Different celebrations of the Hindus, specified in their almanacs are described at length in different works such as the *Tithi Tattva, Tithi Kṛtya, Vratārka, Kālanirṇaya,* the *Kalpadruma* of *Jayasimha, Nirṇaya-Sindhu, Dharmasindhu* and others and also in passages of several *Purāṇas* particularly in the *Bhaviṣyōttara*. The auspicious time for making a gift or *Dāna* and *Vratas* and their results are quoted from various works. Samples of epigraphical evidences are cited in the tables.

Saṃvatsara (Year)

In *Ṛgvēda,* the *Saṃvatsaru* means a year in several passages such as I.110.4; I.140.2; I.161.13; I.164.44; VII, 103. 1, 7, 9; X, 190. 2, we have the derivative form '*Saṃvatśrīṇa*' from *Saṃvatsara.*[1]

In *Ṛgvēda,* X, 62. 2. the word *Parivatsara* was used. *Saṃvatsara* and *Parivatsara* are two of the five names bestowed on the five years of a *Yuga* in the other *Saṃhitas*. In the *Taittirīya Saṃhita*[2] *Namaskāra* is offered to *Rudra* with *Saṃvatsara,* to his bow towards the right with *Parivatsara,* to his bow behind with *idavatsara,* to the bow towards the North with *Iḍāvatsara,* and to the bow above with *Vatsara.* In *Taittirīya Brāhmaṇa,*[3] *Agni, Āditya, Candramas,* and *Vāyu* are identified as *Saṃvatsara, Parivatsara, Iḍāvatsara, Anuvatsara* and it should be noticed that the four names of years are brought in close relation to the four *Cāturmāsyas,* viz., *Vaiśvadēva, Varuṇa, Praghāṣa, Śakamēdha* and *Sunāsīrīya. Kauṭilya* speaks of the *Yuga* of five *Saṃvatsaras* and of the insertion of two *Inter-Calary* months, one at the end of two and half years and the other, at the end of five years.[4]

The *Mahābhārata* knows of the *Yuga* of five years.[5] According to the *Pañcasiddhāntika* of *Varāhamihira, Yuga* means five years of the Sun and the Moon and an Inter-Calary month was added after thirty months.

Regarding the extent of the year in the Vēdic age, in some *Ṛk* verses[6] a year was divided into three, five or six seasons, twelve months, 360 days and 720 days and nights when calculated separately. The *Atharva Vēda*[7] also states that *Saṃvatsara* has twelve spokes and the months have thirty spokes. In the *Brāhmaṇas* also the year is said to have 360 days and 720 days and nights together; *Śatapatha Brāhmaṇa* and *Aitarēya Brāhmaṇa* also say the same.[8] The *Taittirīya Saṃhita* refers to a year of twelve months and also of thirteen months.[9] The *Kauṣītakī Brāhmaṇa* speaks of the thirteenth month[10] with the *Sunasīriya* sacrifice.[11]

The *Jyōtirvidābharaṇa* which is a later fabrication, though it professes that it finished in *Ghaṭakāḷi* 3068 i. e., 33 years before the Christian Era, mentions the names of six persons as the founders of Eras in *Kaliyuga,* viz., *Yudhiṣṭhira, Vikrama, Śālivāhana, Vijayābhinandana, Nāgārjuna, Kalkin*[12] and states that their Eras respectively last for 3044, 135, 18000, 10000, 400000, 821 years. *Aśōka* employs only Regnal Years in his edicts. The fourth pillar edict at *Laurīya-Nandana-Garh* and the fifth pillar edict at *Rāmpūrva* were engraved in the twenty-sixth year of his kingship.[13] *Kauṭilya*[14] also when setting forth the business of the Collector of revenue specifies the times with which the Collector would be concerned as 'The royal year, the month, the fortnight, the day etc.'

Various eras have been used in India for hundreds of years. The lists of eras now in use are met with in inscriptions and Sanskrit literature. Vide preceding chapter.

In the *Saṅkalpa* (declaratory formula) for every religious ceremony, the performer has to mention the grand divisions and sub-divisions of Time beginning with *Śvēta Varāhakalpa* such as *Vaivasvata Manvantara, Kaliyuga* and its first quarter together with the geographical position of the place of performance in *Bharatavarṣa* and the Zodiacal Signs occupied by the Sun, Jupiter and other Planets, the Era, name of year (Jovian), the Month, the Fortnight, the *Tithi, Nakṣatra, Yōga* and *Karaṇa.* A verse of *Dēvala* states that if a man performing a religious rite does not mention the Month, the Fortnight, the *Tithi* and the occasion of the rite, he will not reap the merit of the rite. Such is the significance of eras, years and its sub-divisions in the religious life of our people. Therefore, a calendar *(Pañcāṅga)* is absolutely necessary for every religious Hindu.

At present, the year begins either in *Caitra* or in *Kārtīka* in different parts of India. But in ancient times it began in different months for different purposes. Many *Vēdic* passages show that the reckoning was *Pūrṇimānta,* the year began after the Full Moon of *Phālguṇa* and *Vasanta* was the first season of the year. The *Smṛticandrika* notes that the *Amānta* reckoning is followed in Deccan and *Pūrṇimānta* in North India.[15] *Muṇḍakhēḍe* plate of *Sandraka Jayaśakti* on tenth of *Phālguṇa* dark half of year 602 (must be *Śaka*) is an epigraphical evidence for indicating that in North *Mahārāṣṭra* there was *Amānta* reckoning.[16]

In some inscriptions, the Sixty-year Cycle called *Bārhaspatyamāna* (Jovian) was used. The *Viṣṇudharmōttara*[17] says that the first year called *Prabhava* of Sixty-year Cycle began on *Māgha Śukla,* when the Sun and the Moon occupied the *Nakṣatra Dhaniṣṭha* and were in conjunction with Jupiter. Each year of the *Bārhaspatya* Cycle has the word *'Saṃvatsara'* added to it. The earliest inscriptional references for the *Bārhaspatya* years are two. One to *Siddhārtha-Saṃvatsara* in the fifth year of the reign of the *Cāḷukya* king *Maṅgaḷēśa*[18] and the other is to *Saumya-Saṃvatsara* in the *Ālas* Plate of *Rāṣṭrakūṭa* king *Gōvinda* in *Śaka* 692.[19] Other references can be seen in the tabular statements of this chapter.

In *Bṛhatsaṃhita,*[20] the *phalas* of the sixty years from *Prabhava* to *Kṣaya* are described, as also in *Agni Purāṇa,*[21] while the *Viṣṇudharmōttara Purāṇa*[22] and *Bhaviṣyōttara Purāṇa* quoted in *Jyōtistattva*[23] enumerate all the sixty by name with their *phalas*. How the *Saṃvatsara* names were supposed to indicate different consequences for the years concerned can be illustrated from the *Bṛhatsaṃhita,*[24] when the *Saṃvatsara Prabhava* starts there are drought, strong winds, and destruction by fire and also other calamities, like excessive rain, locusts, etc.; there are diseases due to phlegm but still people are not distressed. The consequences in the next four *Saṃvatsaras* are auspicious, viz., the king rules over a kingdom which is full of paddy, sugarcane, barley and other crops, which is free from dangers in which there are no hatreds, in which the people are joyous and free from the evil results of *Kaliyuga* such as *Adharma,* poverty, accidental deaths etc.

Āyana

The word *'Āyana'* occurs in the sense of 'motion or path' in the *Ṛgvēdȧ.*[25] In the *Śatapatha Brāhmaṇa*[26] the Sun's apparent passage for six months in the North and for six months in the South is referred to, the word *Āyana* is not employed in this passage. *Uttarāyaṇa* and *Dakṣiṇāyana* in later literature mean no more than the Sun's apparent motion or path in the Northern Celestial sphere and in the South for six months is mentioned in *Bṛhadāraṇyaka Upaniṣad* also.[27] *Vasanta* and *Grīṣma* are important parts of *Uttarāyaṇa*, therefore by association with these two and for the sake of symmetry *Varṣa* is also held to be a *Ṛtu* for the Gods.

When the Sun leaves the *Rāśi Dhanus* and enters the *Makara Rāśi* that is called *Makara Saṅkrānti* from which *Udagayana* starts and *Karkaṭa Saṅkrānti* from which *Dakṣiṇāyana* starts. The gifts made on *Uttarāyaṇa* or *Udagayana* yield great reward (merit). A man becomes free from diseases by donating sesame and a draught bull. Sesame should be given to *Brāhmaṇas* on *Uttarāyaṇa Saṅkrānti.* Those who perform *Hōma* with sesame and honour *Brāhmaṇas,* shine in the heavens like those who donate a thousand cows; auspicious lamps fed with oil sesame should be lightened in *Śiva* temples. The installation of images of Gods was suggested in *Uttarāyaṇa*. Images like *Dēvi* can be installed in *Dakṣīṇāyana.* The information on this from epigraphical records is incorporated in the tabular statements of this chapter.

Ṛtu (Season)

The word *'Ṛtu'* occurs several times in *Ṛgvēda*[28] but at one instance we have *'Ṛtūn'* also. The 36 and 37 hymns of second *Maṇḍala* of *Ṛgvēda* are styled *Ṛtavya* hymns. The *Ṛgvēda* itself names five seasons, viz., *Vasanta,*[29] *Grīṣma,*[30] *Prāvṛṣ,*[31] *Śarad,*[32] *Hēmanta.*[33] But the *Ṛgvēda* does not expressly mention *Śiśira*. Three seasons are metaphorically meant in R.V., I.164. 48. and six in R.V., I.164. 15. The *Atharva Vēda*[34] mentions all the six but not in the usual order. The *Aitarēya Brāhmaṇa*[35] says that the year has five *Ṛtus* by putting together *Hēmanta* and *Śiśira*. The *Maitrāyaṇi Saṃhita*[36] speaks of *Saṃvatsara* as having five *Ṛtus* and again as having six *Ṛtus* and the *Śatapatha Brāhmaṇa*[37] says that *Saṃvatsara* comprises six *Ṛtus*. There is mention of seven *Ṛtus* in *Atharva Vēda.*[38] The seventh *Ṛtu* is probably meant to represent the 13th Inter-Calary month as it is expressly mentioned in *Atharva Vēda.*[39] In *Taittirīya Saṃhita*[40] the six *Ṛtus* with two months for each are mentioned. *Vasanta* is mentioned as first of *Ṛtus.*[41]

The *Śatapatha Brāhmaṇa*[42] provides that *Vasanta*, *Grīṣma* and *Varṣa* are the seasons of Gods, *Śarad, Hēmanta* and *Śiśira* are seasons of the *Pitṛs;* similarly the bright half of a month, the day and fore-noon of a day are the times for Gods and the dark half of a month, the night, the after-noon of a day are the times for *Pitṛs* and it winds up with the prescription that a *Brāhmaṇa* should consecrate the sacred fires in *Vasanta,* a *Kṣatriya* in *Grīṣma* and a *Vaiśya* in *Śarad.*[43] As early as the edicts of *Aśōka* the words *Varṣa* (which etymologically means rains) and *'Saṃvatsara'* are both used in the same sense viz., a year in the Brahmagiri Inscription.[44] In some inscriptions, only the *Ṛtu* (Season), the number of the *Pakṣa* and the day occurs; vide *Nāgārjuna Koṇḍa*[45]

Inscriptions (3rd century A.D.), the *Oṃgōḍu* grant of *Vijayaskanda Varman,*[46] the *Nāga* image inscription of *Kaniṣka's* time[47] *Mādharīputa Puriṣadata.*[48] Other references of Seasons can be seen in the tabular statements of this chapter.

Māsa (Month)

The word is either *'Mās'* or *'Māsa'*. We have *Mās* in *Ṛgvēda* I.25. 8; IV. 8. 4; X. 52. 3. He *(Agni)* appears every day and every month and we have *'Māsa'* in *Ṛgvēda* III.31.9; V. 78. 9 and X, 184. 3. *'Mās'* (the measurer) also means the Moon as in *Ṛgvēda,* VIII, 94. 2; X, 12. 7.

In the *Nārada-Saṃhita*[49] it is said that there are nine kinds of reckoning of Time, viz., *Brahma* (of Brahma), *Daiva* (of the Gods), *Mānuṣa* (Human), *Pitṛya* (of the Pitṛs), *Saura, Sāvana, Cāndra, Nākṣatra* and *Bārhaspatya*, but in ordinary worldly matters only five of these are employed. The *Vēdāṅga-Jyōtiṣa* appears to refer to four when it says that in a *Yuga* (of five years) there are 61 *Sāvana* months, 62 Lunar months, and 67 *Nakṣatra* months. *Hēmādri*[50] dwells only three of these reckonings, viz., *Cāndra, Saura* and *Sāvana* while *Mādhava*[51] speaks of two more, viz., *Nākṣatra* and *Bārhaspatya.* The three mentioned by *Hēmādri* are those most employed in religious and civil matters, and *Dharmaśāstra* works are specially concerned with theory while the other two are rarely referred to in the latter.

A *Cāndra* month[52] is a time period from the end of the *Amāvāsya* (New Moon) to the next *Amāvāsya* and twelve such months make the *Cāndra* year of 354 days. It is also called a Lunation. The length of a *Cāndra* month varies from 29. 246 to 29. 817 days owing to the eccentricity of the Moon's orbit and its average period is 29. 53059 days.

A *Saura* month is the time taken by the Sun to move through one Sign of the Zodiac; twelve such months constitute the Solar year[53] and the first day of a Solar year is the first day of the *Saura* month *Mēṣa*. The names of the twelve Solar months distributed two each among the six Seasons are very ancient. They are *Madhu, Mādhava, Śukra, Śuci, Nabhas, Nabhasya, Iṣa, Ūrja, Sahas, Sahasya, Tapas* and *Tapasya.*[54] They occur in inscriptions also as the month *'Sahasya'* and *'Tapasya'* in the *Māndasōr* Inscription of 493 and 529 of the *Māḷava* Era.[55] The name of the *Nakṣatra* is often formed into a derivative and joined to the word *Paurṇamasi, Pūrṇamāsa* or *Amāvāsya* and sometimes the names of the months such as *Phālguṇa* and *Caitra* occur in the *Brāhmaṇas.*

The months ended with the Full Moon *(Pūrṇimānta)* were used by the foreigners like *Kaniṣka* and *Huviṣka* who ruled over Northern India in their records though here and there they employ Macedonian months' names. In the *Zeda* Inscription the Month must be held to be *Pūrṇimānta* as the *Nakṣatra Uttaraphalguṇi* is there said to have occurred on the 20th day of *Āṣāḍha.*[56] The *Tekkali* Plates of *Dēvēndra Varman* record a grant on the occasion of a Lunar eclipse on the 30th day of *Māgha.*[57] Similarly in the *Lodhia* Plates of *Śivagupta Bālārjuna* of *Kōsala* show that the month was *Purṇimānta.*[58] The Inscriptions before the Christian Era rarely mentioned the names of the months such as *Kārtīka* 14th in the *Kharōṣṭhi* Inscription of the Greek king Menoder,[59] 2nd Century A.D. The usual mode is to mention the *Ṛtu* (Season), then the number of the month (and not name) in the Season and then Divasa, e.g., seven Inscriptions[60] from *Mathura* and the *Sārnāth* Inscription of *Mahārāja Kaniṣka.*[61] Sometimes the Month is mentioned but no *Pakṣa,* and the Days are counted continuously from one to thirty, e. g., *Kohi* Plates of *Mahārāja Jayanātha* in year 177 and 22nd day of *Caitra* and *Khoh* Plate of *Sarvanātha* in year 193 month *Caitra* Day 10th without the mention of the Fortnight.[62] Other references are listed in the tabular statements of this Chapter.

The *Śāntiparva* refers to the *Kṣaya* of *Saṃvatsaras, Months, Pakṣas* and *Divasas*.[63] When there is a *Kṣaya Māsa,* the *Adhika Māsa* before it is like all ordinary *Māsas* i.e., there is no prohibition of doing religious acts in it, but the *Adhika Māsa* that follows a *Kṣaya Māsa* is the one that is declared unsuitable for religious rites. The Lunar Month in which no *Saṅkrānti* occurs is called Adhika (Inter-Calary) and bears the same name as that of the next Lunar Month which is called *Śuddha* or *Nija* or *Prākṛta* to distinguish it from the Inter-Calary Month or if in one Solar Month two *Amāvāsyas* occur, then there is an *Adhika* or *Mala Māsa.* The *Kṣaya Māsa* occurs when there are two *Saṅkrāntis* in a Lunar Month, that Lunar Month in which the Sun enters *Mēṣa* Sign is called *Caitra* that in which the Sun enters *Vṛṣabha* is called *Vaiśākha* and so on.

The Inter-Calary Month is called *Adhika Māsa, Mala Māsa, Malimluca, Samsarpa, Amhaspati,* or *Amhahaspati, Puruṣōttama Māsa.* From the very ancient times the thirteenth month was condemned. The *Aitarēya Brāhmaṇa* says that the Gods purchased *Sōma* plant from the thirteenth month and that who sells *Sōma* plant is sinful, that the thirteenth month doesn't exist as a favourable month.[64] In *Taittirīya Saṃhitā* the thirteenth month is called *Samsarpa* and *Amhaspati.*[65] *Amhas* means sin in the *Ṛgvēda.* It is called *Adhika Māsa* because it is an additional month. It is named *Mala Māsa* because this month is as it were the dirt or refuses of time.[66] The word *Malimluca* is found in the *Atharva Vēda.*[67] But its meaning is doubtful. In later literature *Malimluca* means a thief. The Inter-Calary Month is called *Puruṣōttama Māsa* in *Padma Purāṇa.*[68]

The *Agni Purāṇa* provides[69], one should avoid in an Inter-Calary Month the performance of the following acts, viz., consecration of Vēdic fires, the establishment of images, sacrifice, religious gifts, vratas, the recitation of the *Vēda* after undertaking it, *Upanayana,* rites such as the *Saṃskāras* of *Nāmakaraṇa,* inauguration of a king. The general rule is that one should perform obligatory acts and acts prescribed on certain occasions even in an Inter-Calary Month, e. g., daily *Sandhya* worship, the five daily *Mahāyajñas,* the offering into fire etc. Certain matters were to be done only in *Mala Māsa* such as the gift of thirty-three (*Āpūpas* i.e., cakes) every day to *Brāhmaṇas* or at least once in the Inter-Calary Month. Taking interest on money lent, wages of workmen for fixed periods, monthly *Śrāddha,* daily gifts, bath at a holy place, *Japa, Hōma* with barley grains or rice or sesame, rites on the birth of a child, funeral rites, *nava Śrāddhas, Śrāddha* on thirteenth *Tithi* with *Magha Nakṣatra,* gifts and *Japa* on Lunar and Solar eclipse may also be done in *Mala Māsa.*

Pakṣa (Fortnight)

The obvious natural division of Time is the Day and the Lunar Month. A Lunar Month is divided into thirty *Tithis* or Lunar Days of equal mean length. The thirty *Tithis* or Days are divided into two *Pakṣas-Śuddha Pakṣa* or *Śukla Pakṣa* (Bright Half) and *Kṛṣṇa Pakṣa* or *Bahuḷa Pakṣa* (Dark Half) each consisting of fifteen *Tithis*. The fourteen days of each *Pakṣa* are *Prathama* (first), *Dvitīya* (second), *Tṛtīya* (third), *Caturthi* (fourth), *Pañcami* (fifth), *Ṣaṣṭhi* (sixth), *Saptami* (seventh), *Aṣṭami* (eighth), *Navami* (ninth), *Daśami* (tenth), *Ēkādaśi* (eleventh), *Dvādaśi* (twelfth), *Trayōdaśi* (thirteenth), and *Caturdaśi* (fourteenth). The fifteenth or Full Moon day is called *Pūrṇima* (full), and the thirtieth or New Moon day is called *Amāvāsya* (living with Sun). These names are important because a good many Hindu holidays are known by the name of the day and that of the deity in whose honour the day is observed as holy; *Gaṇēśa-Caturthi, Rāma-Navami, Nāga-Pañcami,* etc.

The Lunar Month begins with the Full Moon according to some reckonings and with the New Moon according to others. The former is called *Pūrṇimānta* and the latter is *Amānta.*

The non-mention of the *Pakṣa* and the continuous counting of days from one to thirty continued even in the ninth Century, vide *Kota* Inscription of *Sāmanta Dēvadatta*[70] of *Saṃvat* 879 (923 A.D.) *Māgha Śu.* 20. It may be noted that here the month is *Pūrṇimanta.* In modern times, people use such abbreviations as *Śu Di* or *Va Di* or *Vadya.* The first is abridged from *Śukla Dina* or *Divasa* or *Śuddhạ Dina* and the latter from *Bahuḷa Dina* or *Divasa* (*V*a and *Ba* being often interchangeable). For *'Bahuḷa'* vide Inscription of *Rudrasēna*[71] in the year 127 (held to be of the *Śaka* Era), *Rudradāman's Jūnāgaḍh* Inscription in the year 72 (150 A.D.) of *Mārgaśīrṣa Bahuḷa Pratipada,*[72] *Udayagiri* Cave Inscriptions in Gupta Era 106 (425-26 A.D.) *'Bahuḷa Dinē'.*[73] For *Sūkla* vide the Gupta Inscriptions pp. 95, 102, 114 and for *Śuddha,* vide the *Nāsik* Cave Inscription in the year 42 of the reign of *Uṣavadatta,* son-in-law of *Nahapāṇa* where we have *'Kārtīka Śudhe'*[74] and *Pārḍi* Plates of *Dahrasēna* in *Kālacuri* year 207 (455-56 A.D.). For *'Sū Di'* vide the *Vallabhi* year 106 (525-26 A.D.).[75] For *'Ba Di'* vide J.B.B.R.A.S. vol. I, p. 66; *Bhādrapada Ba* 13 in *Vallabhi* year 210 and E.I., XIX, p. 126 Plate of *Dhṛvasēna* (in *Vallabhi* year 210 *Bhādrapada Ba Di* 9).

It is not clear why the word *'Pakṣa'* is used in many inscriptions before and after the Christian Era even though the word *Pakṣa* occurs in such ancient works as the *Brāhmaṇas*[76] and *Upaniṣads*. The *Candōgya-Upaniṣad*[77] and *Bṛhadāraṇyaka Upaniṣad*[78] employ the words *'Āpūryamāna Pakṣa'* also *'Pūrva Pakṣa'* and *Apara-Pakṣa* in *Bṛhadāraṇyaka Upaniṣad*[79] and *'Apakṣīyamāna Pakṣa'* and *'Apara Pakṣa'* for *'Bahuḷa'* or *'Kṛṣṇa'*. *Pāṇini* derives the word *'Pakṣati'* meaning the first *Tithi* of a *Pakṣa* from the word *'Pakṣa'*, *'Pakṣatih'*.[80]

Tithi (Lunar Day)

The Hindus define a *Tithi* or a Lunar Day, to be the time in which the Moon passes through twelve degrees of her path; and to each *Pakṣa* or half month, they allot fifteen *Tithis,* though they divide the Moon's orb into sixteen phases, named *Kaḷas* one of which they suppose constant, and compare to the string of a necklace or chaplet around which are placed movable gems and flowers.[81] The *Mahākāla* is the day of the conjunction called *Amā* or *Amāvāsya* and defined by *Gōbhila,* the day of the nearest approach to the Sun, on which obsequies are performed to the manes of *Pitṛs* or certain progenitors of the human race, to whom the darker Fortnight is peculiarly sacred.

The word *Tithi* by itself doesn't occur in the *Ṛgvēda* and the other *Vēdic Saṃhitas* but the idea must have been there, *Amāvāsya* is explained in later works as being of two kinds, viz., *Sinīvāli* (day on which *Amāvāsya* is mixed with the fourteenth *Tithi*) and *Kuhū* (when *Amāvāsya* is mixed with the first *Tithi* of the next Fortnight). Similarly, *Paurṇamasi* is of two kinds, *Anumati* (mixed with the fourteenth Tithi) and *Rākā* (mixed with the *Pratipad* of the next *Pakṣa*).

In the *Ṛgvēda, Sinīvāli*[82] is raised to the position of a divinity, is described as the sister of the Gods, offerings are made to her and she is implored to bestow progeny on the worshippers. *Rākā*[83] also is similarly treated in the *Ṛgvēda*. Prayers are offered to *Anumati*[84] to enable the worshippers to see the Sun going up and to bestow happiness and she is spoken of in the same breath with *Sōma, Varuṇa,* and *Bṛhaspati*. The word *Kuhū* occurs in the *Atharva Vēda* and is applied to a deity that is invoked to come to the sacrifice and to bestow on the worshipper wealth and a heroic son.[85] The *Gōbhila Gṛhya Sūtra*[86] when prescribing a day for the setting up of sacred fires mentions the coincidence of an auspicious *Tithi* and *Nakṣatra* and in another place refers to the worship of the Lord of the *Tithi* and of the *Nakṣatra* and of the *Tithi* itself. The *Kauṣītakī Gṛhya Sūtra* speaks of offerings to the *Tithi* on which a child was born and prescribes that the dedication of ponds, wells and tanks was to be made in the Bright Half of a month on an auspicious *Tithi*.[87]

The *Bṛhatsaṃhita* of *Varāhamihira*[88] enumerates about the *Tithipatis* (Lords of Tithis) as follows : 1st *Brahma,* 2nd *Brahma,* 3rd *Hari,* 4th *Yama,* 5th the *Moon,* 6th *Kārtikēya,* 7th *Indra,* 8th *Vasus,* 9th *Nāgas,* 10th *Dharma,* 11th *Śiva,* 12th *Savitṛ,* 13th *Madana,* 14th *Kāḷi,* 15th *Viśvadēvas* and of *Amāvāsya* the *Pitṛs*. He further says that one should do on those *Tithis* the acts that are appropriate to the Lords of those *Tithis*. The *Skanda Purāṇa*[89] states what Tithis are specially dear to who Gods such as eighth and fourteenth to Śiva, fourth to Gaṇēśa, ninth to Caṇḍika. The *Garuḍa Purāṇa*[90] and *Nārada Purāṇa*[91] specify what deities are to be worshipped on the *Tithis* from the first to fifteenth.

Varāhamihira divides *Tithis* into five groups,[92] viz., *Nanda, Bhadra, Vijaya* or *Jaya, Rikta* and *Pūrṇa* and states that on those *Tithis* the acts are appropriate or suited to the Lords of the *Tithis* should be done and become crowned with success and the fruits of actions done on those five classes of *Tithis* yield results similar to their names. The same applies to *Karaṇas* and *Muhūrtas*. The result is *Nanda Tithis* are *Pratipada, Ṣaṣṭhi, Ēkādaśi; Bhadras* are *Dvitīya, Saptami, Dvādaśi; Vijayas* are *Tṛtīya, Aṣṭami, Trayōdaśi; Riktas* are *Caturthi, Navami, Caturdaśi; Pūrṇas* are *Pañcami, Daśami, Pañcadaśi*. Certain combinations of Tithis and certain weekdays yield success and rewards, viz., *Naṇda Tithi* on Friday, *Bhadra* on Wednesday, *Jaya* on Tuesday, *Rikta* on Saturday and *Pūrṇa* on Thursday.

According to *Garga,*[93] *Tithi, Nakṣatra* and *Vāra* (weekday) are means of *Puṇya* (merit) and *Pāpa* (demerit, sin) by being subsidiary to the principal rite prescribed but they are not able to produce the above result independently.

Tithi is of two kinds, *Pūrṇa* (complete or full) and *Sakhaṇḍa*. When a single *Tithi* covers the whole day from Sunrise for sixty *Nāḍikas* (or *Ghaṭikas*) the *Tithi* is *Pūrṇa,* i.e., the *Tithi* begins exactly at Sunrise and ends with the next Sunrise and is exactly sixty *Ghaṭikas* in extent. All other than this are *Sakhaṇḍa,* these latter again are of two kinds, *Śuddha* and *Viddha* (pierced by or intermixed with another).[94] A *Śuddha Tithi* is one that extends from Sunrise to Sunset or up till (in the case of some tithis like *Śivarātri)*. Other *Sakhaṇḍa tithis* are called *Viddha*. *Vēdha* is of two kinds morning *Vēdha* and evening *Vēdha;* the former generally occurs when after six *Ghaṭikas* from Sunrise a *Tithi* touches or gets merged with the next *Tithi,* while the latter (evening *viddha*) generally occurs when a *Tithi* touches another *Tithi* six *Ghaṭikas* before Sunset. In the case of certain Tithis the period prescribed may exceed six *Ghaṭikas*.

Śṛti says,[95] Fore-noon is for the *Gods,* Mid-day for Men and Afternoon is for Pitṛs. *Manu* says,[96] 'one should carry out morning duties, decorating the body, bath, brushing the teeth, applying collyrium to the eyes and worship of Gods in Fore-noon alone'. Therefore all religious acts prescribed for being performed by day on certain *tithis* for Gods must be begun in the morning even if the *Tithi* is mixed with another on that day, but all *Vratas* to be performed in the evening or night have to be performed on the *Tithi* existing in the evening or night even though it may be mixed up *(Viddha)* with another *Tithi*. All rites of the *Śrauta* or *Smārta* type, *Vrata*, *Dāna,* and every other act enjoined by the *Vēda* do not yield proper reward as long as there is no determination of the *Tithi* proper for it.[97] The day for such purposes is divided into five portions between Sunrise and Sunset; 1. *Prātah Kāla,* or early Fore-noon, six *Ghaṭikas* from Sunrise. 2. *Sangava,* or Fore-noon, six to twelve *Ghaṭikas* from Sunrise. 3. *Madhyāhna,* or Mid-day, twelve to eighteen *Ghaṭikas* from Sunrise. 4. *Aparāhna,* or After-noon eighteen to twenty four *Ghaṭikas* from Sunrise. 5. *Sāyāhna,* or late After-noon, twenty four to thirty *Ghaṭikas* from Sunrise. The four *Ghaṭikas* before Sunrise are called *Aruṇōdaya,* or raise of Dawn. The six *Ghaṭikas* after sunset are called *Pradōṣa,* or Evening. The two *Ghaṭikas* in the middle of night are called *Niśīdha,* or Mid-night. A festival marked as *Pūrva Viddha* is celebrated on the first day of the *Tithi,* not on second. A festival celebrated on the second day on which the *Tithi* is current is said to be *Para Viddha.* When one *Tithi* commencing and the other ending between eighteen and twenty-four *Ghaṭikas* after Sunrise, when a similar meeting doesn't take place next day is called *Tithidvayam.* In some works, it is stated that when the Full Moon, which is in the Lunar mansion that gives the name to a month, is in the conjunction with Jupiter that *Tithi* has the word *Mahā* is prefixed to it. For example, the

Paurṇima of *Kārtīka* would be called *Mahā Kārtīki,* if the Moon and Jupiter are both in the *Kṛttikas* on that *Tithi*. The *Rājamārtāṇḍa* and *Bhaviṣya Purāṇa* provide that a bath on the *Mahā Caitri* and on the remaining eleven *Paurṇimas* with prefix *Mahā* in certain holy places yields great rewards, viz., bath at *Prayāga* on *Mahāmāghi,* on *Mahā Phālguṇi* in the *Naimiśa* forest, on *Mahā Caitri* at *Sālagrāma,* on *Mahā Vaiśākhi* at *Mahādvāra,* on *Mahā Jyaiṣṭhi* at *Puruṣōttama,* on *Mahāṣāḍi* at *Kanakhala,* on *Mahā Śrāvaṇi* at *Kēdāra,* on *Mahā Bhadri* at *Badari,* on *Mahāśvini* at *Kubjāmra,* on *Mahākārtīki* at *Puṣkara,* on *Mahā Mārgaśīrṣi* at *Kānyakubja,* on *Mahāpauṣi* at *Ayōdhya.*[98]

Vāra (Week day)

One of the five *Aṃgas* in the Indian *Pañcāṅgas* is the Weekday. The interval between two Sunrises was considered as the most important unit of Time. This is Solar and is called the Civil day. But *Tithi* is a Lunar division of Time and begins and ends at various places in the Solar day. The names of the days of the week bear closer affinity to Greek names than English names do. The present order of weekdays appears to be planetary, but if we look to the mere names, the planets are arranged as Sun, Moon, Mars, Mercury, Jupiter, Venus and Saturn. *Ravi Vāra* is the first day in the week for various reasons, one of which is that creation is believed to have begun on that day. Sunday is called *Ravivāra*, Monday-*Sōmavāra,* Tuesday-*Maṅgaḷavāra*, Wednesday-*Budhavāra,* Thursday-*Bṛhaspativāra,* Friday-*Śukravāra,* Saturday-*Śanivāra*. There are elaborate astrological treatises dealing with the influence of the planets on the destiny of individuals on each day, and what must and must not be done everday of the week. A Solar day or *Divasa* is divided into sixty *Ghaṭikas,* each *Ghaṭika* is being equivalent to twenty-four minutes. A *Ghaṭika* is subdivided into sixty *Palas,* a

Pala into sixty *Vipalas* and a *Vipala* into sixty *Prativipalas*. The *Prativipala* is the smallest Hindu division of Time and is equivalent to about 0.006 seconds.

It appears that the continuous seven-day week did not form part of the religious life of any nation in antiquity and even now it is no part of the religious life of many people.

The oldest inscriptional evidence so far available is that of the *Erān* Stone Inscription of *Budha Gupta* in the *Gupta* year[99] 165 (484 A.D.) that mentions Thursday and the twelfth *Tithi* and a Bright Half of *Āṣāḍha*.

In *Bṛhatsaṃhita* occurs this 'when *Pitāmaha* declared that the weekday Tuesday doesn't lead to auspicious results or when (I declare) that the day of the son of the Earth is inauspicious, what difference is there between (the *Śāstra*) composed by a divine author and that by a human author (like myself)?[100] The same book sets out in chapter 103 (verses 61-63) what actions are proper to be done on the weekdays from Sunday to Saturday.

In *Vaikhānasa Smārta Sūtra* (I, 4) and *Bauddhāyana Dharma Sūtra* (II, 5. 23) *Tarpaṇa* (satiation with water) is offered to the Sun, the Moon, Mars, Mercury, Jupiter, Venus, Saturn and to Rāhu and Kētu and the former mentions *Budhavāra* (Wednesday) in II. 12.

The *Atharvaṇa-Vēdāṅga Jyōtiṣa* points out what should be done on each of the seven weekdays from Sunday to Saturday.[101] The *Nārada Purāṇa* provides for a sacrifice to nine grahas.[102] The *Matsya Purāṇa*[103] provides for more elaborate sacrifice to planets. The same *Purāṇa* provides that all weekdays except Sunday and Tuesday are auspicious.[104] The *Kūrma Purāṇa*[105] provides that *Śrāddhas* on weekdays from Sunday to Saturday respectively yield the benefits of health, happiness or prosperity, success, all desires, desired learning, wealth and long life.

Nakṣatra (Constellation)

The word *'Nakṣatra'* has in all three senses, (1). Star in general, (2). Twenty-seven equal parts of the Zodiac, (3). Asterisms in the Zodiacal belt (which may each consists of one or more Stars). The first the third are the most frequent meanings of the word *Vedic Saṃhitās*. The word *Nakṣatra* occurs frequently in the *Ṛgvēda* and the other *Saṃhitas* and *Brāhmaṇas*. 'The *Nakṣatras* like thieves go away along with nights to make room for the Sun that sees the world'.[106] But in *Ṛgvēda* X, 85. 2. and in X, 68. 11. *Nakṣatra* appears to mean the twenty-seven well-known asterisms. The word *'Ṛkṣa'* in the sense of Star occurs in the *Ṛgvēda* 'these *Ṛkṣas*, that are established high up in the sky are seen at night but where did they go by day'.[107]

The *Nakṣatras* are 27 or 28 adding Abhijit after *Uttarāṣāḍha* and before *Śravaṇa* in ancient authorities. In the *Vēdic* literature, *Vēdāṅga Jyōtiṣa* and even in *Yājñavalkya-Smṛti* they are enumerated from *Kṛttika* to *Apabharaṇi* (or *Bharaṇi*) while in works from the 3rd or 4th Century A.D. and in modern times they are enumerated from *Aśvini* to *Rēvati*. In the *Atharvaṇa Nakṣatra Kalpa*[108] the *Dēvatas* (deities) of the *Nakṣatras* are given. In the same book chapter II states the number of stars in each *Nakṣatra*. In some *Purāṇas* also such as the *Viṣṇudharmōttara* (I, 83. 13-21) the presiding deities of *Nakṣatras* are set out. *Bṛhatsaṃhita,*[109] specifies the deities of *Nakṣatras* from *Aśvini* to *Rēvati* including *Abhijit*.

The *Nakṣatras* were closely concerned not merely in a religious rite called *Nakṣatrēṣṭi,* but they were of prime importance in the basic *Śrauta* rite, viz., consecration of sacred fires *(Agnyādhāna)*. In the *Śatapatha Brāhmaṇa*[110] reference is made to several *Nakṣatras* from *Kṛttika* onwards together with their presiding deities as being fit for *Agnyādhāna*.

The *Nakṣatras* had very early been classified as *Puṇya* (auspicious) and *Pāpa* (inauspicious) and into male and female.[111] The *Vēdāṅga Jyōtiṣa (Yājuṣa)* gives a classification of *Nakṣatras* into *Ugra* and *Krūra*.[112] The *Nakṣatras* are further classified in the *Bṛhatsaṃhita,* (Chapter 97, 6-11) into *Dhruva* (or *Sthira*=fixed), *Tīkṣaṇa* (or *Dāruṇa*), *Ugra* (or *Krūra), Kṣipra* (or *Laghu*), *Mṛdu* (or *Maitra*), *Mṛdu Tīkṣaṇa* (or *Sādhāraṇa* or *Miśra*), *Cara* (or *Cala*).[113] The *Bṛhatsaṃhita,* states that on *Dhruva Nakṣatras* the coronation of a king, *Śānti* (propitiatory rite for averting impending evil or calamity), planting of trees, foundation of a city, *charitable* acts, sowing seeds and other permanent acts should be done; on *Tīkṣaṇa Nakṣatras* succeed attempts to harm, the acquisition of *mantra* (mystic verse or formulae), raising ghosts, arrest (of a person), beating, separating (two friends or) relationships; *Ugra Nakṣatras* are to be used for success in extermination, destruction of another's property, cheating, arrest, poisoning, incendiarism, striking with a weapon, killing; *Kṣipra* (or *Laghu*) *Nakṣatras* are declared as leading to success in sales, in making love, in acquisition of knowledge, decoration, arts, crafts such as carpentry, medicines, journeys; the *Mṛdu Nakṣ atras* are beneficial in securing friends, sexual affairs, clothes, ornaments, auspicious ceremonies (Marriage, Upanayana, etc.) and singing; the *Mṛdu-Tīkṣaṇa* (or *Sādhāraṇa*) *Nakṣ atras* produce mixed results (on which mild or fierce acts may be done); the *Cara Nakṣatras* are beneficial in doing fleeting actions. The *Muhūrtamārtāṇḍa* provides that wise men engage in actions simliar to the names of the groups of *Nakṣatras* for attaining success.[114] It should be noticed that some works like *Jyōtiṣa Ratnamāla* (III, 9)and the *Muhūrta Cintāmaṇi* (II, 2-8) hold that Sunday, Monday, Tuesday, Wednesday, Thursday, Friday and Saturday are respectively *Dhruva, Cara, Ugra, Miśra, Laghu, Mṛdu* and *Tīkṣaṇa* and that actions that are appropriate for *Nakṣatras* of those groups are appropriately performed on respective weekdays.

The *Jyōtiṣa Ratnamāla* divides twenty eight *Nakṣatras* (including *Abhijit*) into four groups of seven each, called *'Aṃdhākṣa'* (blind), *'Mandākṣa'* (of dim sight), *'Madhyākṣa'* (of medium sight) and *'Svakṣa'* (of good sight) and states that property stolen on *'Aṃdhākṣa' Nakṣatras* may be recovered quickly, stolen on *'Mandākṣa' Nakṣatras* after effort, stolen on *'Madhyākṣa' Nakṣatras* there is no recovery but the owner may hear that it has been taken to a distant place by the thief, stolen on *'Svakṣa'* or *'Sulōcana' Nakṣatras* the owner would not recover it nor even hear about its whereabouts.

It would be noticed that a *Nakṣatra Astrology* apart from the *Rāśi Astrology* had been thoroughly developed in India. The twelve *Rāśis* are to be found in the Zodiacal belt. Each extends over 2¼ *Nakṣatras* i. e., *Mēṣa* extends over *Aśvini, Bharaṇi* and ¼th of *Kṛttika* and so on. The donations made on various *Nakṣatras* can be seen in the tabular statements of this Chapter.

Yōga (Conjunction)

The fourth *Aṃga* in a *Pañcāṅga* is called *Yōga*. There is no direct astronomical phenomena corresponding to it. A *Yōga* is the Time during which the Sun and the Moon together accomplish 13 degrees 20 minutes of space. There are thus 27 *Yōgas* which together make 360 degrees as stated in the *Ratnamāla* IV, 1-3. and are as follows.

S. No	Name	Deity	S. No	Name	Deity
1	Viṣkambha	Yama	15	Vajra	Varuṇa
2	Prīti	Viṣṇu	16	Siddhi	Gaṇēśa
3	Āyuṣmān	candra	17	Vyatīpāta	Śiva
4	Saubhāgya	Brahma	18	Varīyān	Kubēra
5	Śōbhana	Bṛhaspati	19	Parigha	Viśvakarman
6	Atigaṇḍa	Candra	20	Śiva	Mitra
7	Sukarman	Indra	21	Siddha	Kārtikēya
8	Dhṛti	Āpah	22	Śādhya	Sāvitri
9	Śūla	Sarpa	23	Śubha	Kamala
10	Gaṇḍa	Agni	24	Śukla	Gauri
11	Vṛddhi	Sūrya	25	Brahman	Aśvinau
12	Dhṛva	Pṛthvi	26	Aindra	Pitṛs
13	Vyāghāta	Pavana	27	Vaidhṛti	Aditi
14	Harṣaṇa	Rudra			

There are *Nitya* (ever recurring) *Yōgas*. The *Ratnamāla* states that they yield good or bad results in keeping with their names. The *Muhūrta Darśana*[115] states that among these twenty seven *Yōgas,* nine are condemned viz., *Parigha, Vyatīpāta, Vajra, Vyāghāta, Vaidhṛti, Viṣkambha, Śūla, Gaṇḍa* and *Atigaṇḍa*.[116] The *Ratnamāla* provides that *Vyatīpāta* and *Vaidhṛti* are wholly inauspicious. The first half of *Parigha* is so and only the first quarter of those *Yōgas,* the names of which are odious; and that one should avoid in all auspicious rites the first three *Ghaṭikas* in *Viṣkambha* and *Vajra,* nine *Ghaṭis* in *Vyāghāta,* five in *Śūlā,* six in *Gaṇḍa* and *Atigaṇḍa*. The *Agni Purāṇa* has similar verses.[117] The *Kālanirṇayakārika* provides that when a *Yōga* spreads over two days, fasts and the like are to be observed on the first of the two days provided that *Yōga* exists at Sunsct on that day and *Vratas* and religious gifts are to be performed on the latter of two days provided the *Yōga* exists for three *Muhūrtas* (i.e., 6 *Ghaṭis*) after Sunrise and for *Śrāddha* the day on which the *Yōga* extends over the time required for the *Śrāddha* rite should be accepted. [118] The system of *Yōgas* must be held to be ancient. The *Yājñavalkya Smṛti* (I.218) when specifying the times for the performance of *Śrāddhas* includes *Vyatīpāta* among them. In the *Harṣacarita* of *Bāṇa* it is said (Ucchavās-IV) that *Harṣa* was born on a day free from the taint of all defects such as *Vyatīpāta (Vyatīpātādi-Sarvadōṣābhisaṅga-Rahitēkṣaṇ)*. There would generally be thirteen *Vyatīpātas* in a year (sometime 14) and among the 96 *Śrāddhas*, *Śrāddhas* on these thirteen *Vyatīpātas* are included. [119]

Besides these 27 *Yōgas,* there are others that occur when there are special conjunctions of weekdays with certain *Tithis* and *Nakṣatras* or when certain Planets occupy certain *Rāśis* on certain *Tithis* and *Nakṣatras.* Apart from inscriptions in the *Vikrama Era,* where *Yōgas* are freely cited, one hears of the *Yōga* proper in such auspicious combinations as the following [120] —*Kapilaṣaṣṭhi Yōga,* i.e., the combination of *Bhādrapada Kṛṣṇa-Ṣaṣṭi* (in a year in which either *Bhādrapada* or some previous Lunar month is *Adhika*) with *Nakṣatra Rōhiṇi* and *Yōga Vyatīpāta* on a Tuesday.

Ardhōdaya Yōga, i.e., a combination on a Sunday by daytime, of *Nakṣatra Rōhiṇi, Yōga Vyatīpāta,* and *Māgha Amāvāsya* (i.e., the *Amāvāsya* at the end of Lunar *Māgha*).

Campā Ṣaṣṭhi, i.e., a combination on a Sunday or Tuesday of *Mārgaśira Śukla Ṣaṣṭhi* with *Nakṣatra Śatabhiṣa* and *Yōga Vaidhṛti.*

Mahāmahā Vāruṇi, i.e., a combination of *Phālguṇa Kṛṣṇa Trayōdaśi* with Saturday and *Nakṣatra Śatabhiṣa* and *Yōga Śubha.*

Daśahari, i.e., a combination in the Fore-Noon of *Jyēṣṭha Śukla Daśami* with Wednesday or Tuesday and *Nakṣatra Hasta* and *Yōga Vyatīpāta.*

Mahāśivarātri, i.e., *Māgha Kṛṣṇa Caturdaśi* combining at Midnight with a Sunday or Tuesday and *Yōga Śiva.*

Vyatīpāta itself has two technical senses apart from being 17th out of the 27 *Yōgas* mentioned above viz., (1) when *Amāvāsya*[121] occurs on a Sunday and the Moon is in the 1st quarter on any one of the *Nakṣatras Śravaṇa, Aśvina Dhaniṣ ṭha, Ārdra,* and *Āślēṣa* i.e., *Vyatīpāta* (2) when on the 12th of the Bright Half Jupiter and Mars are in the Sign Lion, the Sun is in Ram *(Mēṣa)* and when that *Tithi* has *Hasta Nakṣatra* on it, it is *Vyatīpāta Yōga.*[122] These two are sometimes called *Mahāvyatīpāta*. Gifts made on those conjunctions are very much commended. Gifts on *Amāvāsya* yield hundred-fold merit, on the suppression of a *tithi,* a thousand-fold, on *Viṣuva* day ten thousand-fold but on *Vyatīpāta* endless merit. By gifts of cows, land, gold and clothes made on this, the donor becomes free from sin and secures God-head, the position of Indra, health, the rulership of men.[123]

The *Sūrya Siddhānta* gives an explanation on *Vyatīpāta* and *Vaidhṛti*—when the Sun and the Moon are on the same side of either Solstice and when the Sun of their longitudes being a circle they are of equal declination, it is styled *Vaidhṛti.* When the Moon and the Sun are on opposite sides of either Solstice and their minutes of declination are the same it is *Vyatīpāta,* the Sun of their longitudes being a half circle.[124] The twenty seventh *Yōga* called *Vaidhṛti* is just like *Vyatīpāta* in all respects. *Bharadvāja* says that the merit of gifts made on *Vyatīpāta* and *Vaidhṛti* is endless.

In the *Pañcāṅgas* several *Yōgas* like *Amṛtasiddhi, Yamagaṇḍa, Dagdha Yōga, Mṛtyu Yōga, Ghabāḍa* are mentioned. They are purely astrological. When on Sunday there is *Hasta* on Monday *Mṛgaśira* on Tuesday *Aśvini*, on Wednesday *Anūrādha,* on Thursday *Puṣya,* on Firday *Rēvati,* and on Saturday *Rōhiṇi,* all these are *Amṛta Yōgas* (Fortunate ones). *Anūrādha* on Sunday, *Uttarāṣāḍha* on Monday, *Śatatāraka* on Tuesday, *Aśvini* on Wednesday, *Mṛgaśīrṣa* on Thursday, *Āślēṣa* on Friday, and *Hasta* on Saturday—these are seven *Mṛtyu Yōgas* (inauspicious). If the week-days from Sunday to Saturday occur on the *tithis* 12th, 11th, 10th, 3rd, 6th, 2nd, 7th respecitively, these are *Dagdha Yōgas*.

Karaṇa

The word *Karaṇa* is derived from *Kṛ* (to do) and said to have been so named, because it divides a *Tithi* into two parts.[125] *Karaṇa* is the fifth item in the *Pañcāṅga*. As the *Karaṇa* is the half of a *Tithi,* there are 60 *Karaṇas* in Lunar month. *Karaṇas* are of two kinds viz., *Cara* (moving) and *Sthira* (immovable). According to *Bṛhatsaṃhita* (Chap. 99, 1-2) the seven *Cara Karaṇas* with their presiding deities are:

(1) Bava-Indra
(2) Bālava-Brahma
(3) Kaulava-Mitra
(4) Taitila-Aryama
(5) Gara (or Garaja) Earth
(6) Vaṇija-Śrī
(7) Viṣṭi-Yama

The four *Sthira Karaṇas* with their deities are:

(1) Śakuni-Kāḷi
(2) Catuṣpada-Vṛṣa
(3) Nāga-Snake
(4) Kiṃstughna-Vāyu

This division of *tithis* into two halves resembles the division of *Rāśis* into two *Hōras*.[126] The *Sthira Karaṇas* occur only once in a month i. e., on the latter half of 14th *Tithi* of dark half occurs *Śakuni; Catuṣpada* and *Nāga* occupy the two halves of *Amāvāsya* and *Kimstughna* occurs on the first half of *Pratipada,* i.e., if the thirty *tithis* of an *Amānta* month are divided into 60 halves, *Śakuni, Catuṣpada* and *Nāga* occupy the 58th, 59th, and 60th parts and *Kimstughna* occupies the first half of the *Pratipada* of the next month. *Bava* occupies the 2nd half of *Śukla Pratipada,* and the others follow one after another and as they are only seven, they are repeated eight times in a month from the second half of the *Pratipada* to the first half of the 14th *Tithi* of Dark Half which would be 57th of the 60 halves of the *tithis* of the month and would be occupied by *Viṣṭi* in the *Pañcāṅgas,* two *Karaṇas* should properly be shown under each *Tithi,* but in *Pañcāṅgas,* only the *Karaṇa* is shown against a *Tithi* whose ending movement is thirty *Ghaṭikas* or less from Sunrise.

The *Karaṇas* are only of astrological use and must have been named many Centuries before 400 A.D. The *Nārada Purāṇa* mentions both kinds of *Karaṇas*.[127] The *Bṛhatsaṃhita*, lays down what should be done or not to be done on the several *Karaṇas*,[128] as follows—One should perform on *Bava* all auspicious (religious) rites, all acts that bring about physical health; on *Bālava* religious acts and acts beneficial to *Brāhmaṇas;* on *Kaulava* acts leading to good relations and friendship and choosing a bride; on *Taitila* acts that lead to popularity with people, to securing the patronage of some important person and building a house; on *Gara* actions pertaining to ploughing sowing seeds and those that arise in houses; on *Vaṇija* acts of a permanent nature, trading and meeting with people; whatever is done on *Viṣṭi* doesn't bring about fortunate results; but actions as regards destruction of enemies, poisoning and incendiarism become successful on it; on *Śakuni* one should do what leads to Physical health and Meditation, Planting roots and eating them and becoming successful in Mantras; on *Catuṣpada*[129] actions relating to cows (i. e., gifts of them, and rearing them), actions regarding *Brāhmaṇas* and one's ancestors and king's business should be done; on *Nāga* acts relating to immovable property, cruel acts depriving a man of his wealth or wife and acts that lead to hatred among people; on *Kimstughna* auspicious (religious) acts, *Vēdic* sacrifices and acts leading to physical prosperity and *Maṅgaḷa* acts (marriage etc.) and other acts that bring success in these matters.

In the Dark Half *Viṣṭi* will occur in the latter half of the third and tenth *tithis* and in the 1st half of the seventh and fourteenth *tithis;* in the Bright Half *Viṣṭi* will occur in the 4th and 11th *tithis* and the 1st half of the 8th and 15th (i.e., *Pūrṇamasi) tithis*. Great sages declare that *Viṣṭi* doesn't confer fortunate results, therefore a good man should avoid this.

Muhūrta

The word *'Muhūrta'* means 'a short time, a few moments'. This *Muhūrta* occurs twice in the Ṛgvēda.[130] The meaning of *'Muhūrta'* is also found in the *Śatapatha Brāhmaṇa.*[131] There is another meaning of the word 'Muhūrta' in *Śatapatha Brāhmaṇa,* X, 4.2.18 and XII, 3.2.5 where it is said that there are fifteen *Muhūrtas* of the day and fifteen of the nights, i.e., 30 in *Ahōrātra*, and that in a year there are 10800 *Muhūrtas* (30×360).[132] Here the word *'Muhūrta'* means fifteenth part of the day, i.e., in a general way about two *Nāḍikas* or *Ghaṭikas.* The fifteen names of the *Muhūrtas* of the day and fifteen *Muhūrtas* of the night occur in *Taittirīya Brāhmaṇa*[133], III, 10. 1. 1-3.

The *Bauddhāyana Dharmasūtra,* II,10.26. Manu IV, 92. and *Yajñavalkya Smṛti*-I, 115 prescribe that a house holder should get up from his bed in the *Brahma Muhūrta.*[134] The *Brahma Muhūrta* is mentioned in *Drōṇaparva.*[135] Besides, in several places auspicious *Tithis, Nakṣatras* and *Muhūrtas* are mentioned in general. The *Muhūrta Darśana* of *Vidya-Mādhavīya* gives names of 15 *Muhūrtas* and states that seven of these are auspicious, viz., *Abhijit, Vairāja, Śvēta, Sāvitra, Maitra, Bala, Vijaya.*[136] In *Udyōgaparva,* it is stated that a *Purōhita* was sent as *Dūta* of the *Pāṇḍavas* on *Puṣya-Yōga* and *Jaya Muhūrta.*[137] The *Vāyu Purāṇa* enumerates the names of the fifteen *Muhūrtas* in the day some what differently and also of the night.[138] The *Matsya Purāṇa* (in Chap. 22 2) refers to two *Muhūrtas, Abhijit* and *Rauhiṇa* and mentions eight *Muhūrtas* as auspicious on which to begin the construction of a new house.[139] *Varāhamihira* remarks : Whatever has been stated as proper for being done on certain *Nakṣatras,* may be done on *Tithis* presided over by the deities of those *Nakṣatras* and also on *Karaṇa* and on *Muhūrtas;* that leads to success as the deity is the same.[140] Some examples may be cited here : On *Raudra* may be done whatever is of a terrific nature; on *Maitra* should be done whatever is affectionate or friendly; on *Sārabhaṭa* black magic may be resorted to against enemies; *Abhijit* is proper for all desired objects and gives success in all undertakings; *Vijaya* leads to victory if one marches on it, one may perform auspicious acts on it and *Śānti* rites. As some *Muhūrtas* of the day of the duration of two *Ghaṭikas* were declared to be auspicious, gradually a third meaning came to be attached to this word, viz., 'time that is fit for the performance of an auspicious act'.[141] Almost all the works on *Kāla* discuss the proper times for performing the *Saṃskāras* and religious rites. For example, *Garbhādāna, Pumsavana, Jātakakarma, Caula, Upanayana,* topics connected with marriage; times for kindling sacred domestic fire, times for construction of a house and for first entrance into it, the proper times for starting on a journey or marching against an enemy king; auspicious and inauspicious *Śakunas* (Omens or Prognostications); times for king's coronation, for wearing of rich clothes and ornaments and for agricultural operations, for sale and purchase of animals, for bath with sesame and myrobalan, for finding lost articles, for constructing wells and tanks, the results of a lizard falling on one's body, what planet is auspicious or inauspicious in which sign from other sign of ones birth, the *Puṇyakāla* of *Saṅkrāntis* etc. The proper times for the establishment of the images of Gods are specified in many works such as *Bṛhatsaṃhita, Matsya Purāṇa, Viṣṇu Dharmōttara,* Hemādri on *Kāla, Nirṇaya-sindhu, Dharmasindhu.* The *Bṛhatsaṃhita*[142] lays down the general rules for the establishment of the images of all Gods in the Northward passage of the Sun in the Bright Half, when the Moon is in the *Varga* of Jupiter, when the *Lagna* is a *Sthira Rāśi* and the *Navaṃśa Rāśi* of the *Lagna* is *Sthira* and beneficent planets occupy the *Kēndra* places or the fifth and ninth places in the horoscope of the establisher, when the malefic planets are in the third, sixth, tenth or eleventh places, on *Nakṣatras* called *Dhṛva* or *Mṛdu* or on *Śravaṇa, Puṣya* or *Svāti Nakṣatra* on a favourable week day (other than Tuesday) the establishment of Gods is commended.

The *Ratnamāla*[143] provides different *Nakṣatras* for the establishment of the images of different Gods, one interesting provision being that the image of Buddha was to be established on *Śravaṇa Nakṣatra.* In the case of the images of the *Mātṛs, Bhairava,* the *Varāha, Narasimha* and *Trivikrama* incarnations and of *Dēvi,* the killer of *Mahiṣāsura,* establishment may be made even in *Dakṣiṇāyana.* For the establishment of *Liṅga* special rules are provided by *Nirṇayasindhu.*

In the Indian Antiquary Vol. VI, p. 363 we find an inscription of *Cāḷukya* king *Maṅgalīśa* date *Śaka* 500 (578-79 A.D.) recording the grant of a village named *Lāñjīśvara* on the auspicious occasion of the establishment of an image of God *Viṣṇu.*

Some inscriptions show that kings rewarded men for astrological devices. For example, the *Kadamba* Plates of *Rāṣṭrakūṭa* kings *Prabhūtavarṣa Gōvindarāja*-III granted in *Śaka* 735 to a *Jain muni* the village of *Jalamaṅgaḷa* for having warded off the evil influence of Saturn from *Vimalāditya* one of his Governors.[144]

Grahaṇa (Eclipse)

Great importance was attached from very ancient times to Eclipses. An extensive literature has accumulated on the subject. *Hēmādri* on *Kāla* 379-394, *Kālavivēka* 521-543, *Kṛtyaratnākara* 625-631, *Kālanirṇaya* 346-358, *Varṣakriyā-kaumudi* 90-117, *Tithitattva* 150-162, *Kṛtyatattva* 432-434, *Nirṇayasindhu* 61-76, *Smṛtikaustubha* 69-80, *Dharmasindhu* 32-35, *Gadādharapaddhati (Kālasāra)* 588-599 are the important among them. A total Eclipse of the Sun is referred to even in the *Ṛgvēda* V, 40. 5, 6, 8. 'O Sun! when *Svarbhānu* the son of an *Asura* struck you with darkness, O Indra! when you struck down the wiles of *Svarbhānu* that existed below the Heaven, *Atri* secured the Sun, which had been concealed by darkness that deviated from the usual role (or law) by means of the highest prayer'.[145] It is stated in the *Sāṅkhyāyana Brāhmaṇa*[146] that the *Atris* performed the *Saptadaśa-Stōma* three days before *Viṣuvat* and thereby smote the demon *Svarbhānu* that had pierced the Sun with darkness, i.e., the Solar Eclipse took place three days before the autumnal *Viṣuvat* (Equinox).

The real cause of Eclipses was known to Indian astronomers several Centuries before *Varāhamihira* who remarks :[147] 'In a Lunar Eclipse the Moon enters into the shadow of the Earth and in a Solar Eclipse the Moon enters the Sun (i.e., comes between the Sun and the Earth) that this cause of Eclipses was found by former *Ācāryās* who had a divine insight; that *Rāhu* is not the cause (of Eclipse); this is the true state that the *Śāstra* declares'. In spite of this correct theory of Eclipses, common people and even learned men then believed and do believe up to the present day that *Rāhu* is the cause of Eclipses and regard an Eclipse as a special occasion for bath, gifts, *Japa, Śrāddha* and the like. According to *Vyāsa*[148] an Eclipse of the Moon is one hundred thousand times (more meritorious than an ordinary day) and an Eclipse of the Sun ten times more so than the preceding and if the waters of the *Ganges* are near (for a bath) then a Moon's Eclipse is one crore of times (more meritorious) and a Sun's Eclipse is ten times more so than the preceding'.

The person on seeing an Eclipse is to bathe. It is said[149], on seeing *Rāhu* (i.e., Eclipse) men of all *Varṇas* become impure. They should first bathe and then do their usual duties and should give up food already cooked (before the Eclipse). Peculiar sanctity was deemed to attach to the time of *Grahaṇa.* If a man doesn't take a bath at the time of Eclipse, at the passage of Sun from one *Rāśi* to another, there is no doubt that he would suffer from leprosy for seven future lives and sorrow would be his lot.[150] He should take a bath in cold water and in as holy a place as possible. The holiest bath is the one in the *Ganges* or *Gōdāvari* or at *Prayāga,* then in any one of the big rivers such as the six rivers connected with *Himavat* mountain and the six South of the *Vindhya* mentioned in the *Brahma Purāṇa,* then in any other water, since at the time of Eclipse all water becomes holy like *Ganges* water.[151] A bath with hot water was allowed only to children, old men and persons that were ill.

One had to take a bath when the Eclipse began, to perform *Hōma,* worship of Gods and *Śrāddha* while the Eclipse was in progress to take gifts when the Eclipse was about to end and to take a bath again when the Sun or Moon became free from Eclipse.[152] Even a person who is impure owing to a birth or death has to take a bath on an Eclipse but he is not to make a gift or to perform *Śrāddha,*[153] but the *Madanaratna* and *Nirṇayasindhu* refute this view and hold that in an Eclipse even one who is in *Aśauca* can take a bath, perform *Śrāddha* and *Puraścaraṇa.*[154] In the *Purāṇas* and Medieval digests some distinction as regards merit or holiness was made between Eclipses in certain months and baths in certain rivers or holy places. The *Kālanirṇaya* especially commends the *Gōdāvari* for a bath in a Lunar Eclipse and one in the *Narmada* for a Solar Eclipse. The *Kṛtya Kalpataru (Naiyataka), Hēmādri* on *Kāla* and the *Kālavivēka* quote a long passage from the *Dēvī Purāṇa,* 'an Eclipse in *Kārtīka* is the highest in merit at the confluence of the *Ganga* and *Yamuna,* in *Margaśīrṣa* on *Dēvika,* in *Pauṣa Narmada* is holy, in *Māgha Sannihita* is holy' etc.

The general rule is that one should not take a bath, make a gift or perform a *Śrāddha* at night. *Āpastambha* says[156] 'let him avoid a bath of night'. *Manu* says [157] 'one should not perform a *Śrāddha* at night; she (night) is declared to be a *Rākṣasi* (a demon) and also at both Twilights and when Sun has risen'. But Eclipses were an exception as regards bath, gift and *Śrāddha. Yājñavalkya* enumerates Eclipses among the proper times for *Śrāddha.*[158] It is stated by *Śatātapa* that *Gifts, Tapas* and *Śrāddha* at the time of Eclipses yield inexhaustible rewards or merit; the Night is a *Rākṣasi* elsewhere (at times other than those of Eclipses); therefore one should avoid it in other matters.[159] A passage is quoted from the *Mahābhārata :* 'on *Āyana* and *Viṣuva* days and on the occasion of the Eclipses of the Sun and the Moon one should make a gift of land together with *Dakṣiṇa* to a deserving *Brāhmaṇa'. Yājñavalkya* succinctly observes, 'not by learning alone nor by tapas alone does one become a deserving person; that is declared a deserving person *(Pātra)* in whom these two *(Vidya)* and *(Tapas)* and actions appropriate to these are found'.

Numerous inscriptions published so far bear witness to the fact that over the whole of India, ancient and medieval kings and well to do people carried out to the latter this recommendation of making gifts of land on Eclipses. Some inscriptions are mentioned here by way of illustration. Vide for gifts on Solar Eclipses the following:

I.A., Vol. VI, pp. 72-75 (in *Śaka* 534 expired) the grant of a village to a *Brāhmaṇa* of *Tagara* by the great king *Satyāśraya Pulakēśi*-II on *Bhādrapada Amāvāsya;* E.I., Vol. III, pp. 1-7; the *Paṭṭaḍakal Pillar Inscription* of Western *Cāḷukya Kīrtivarman*-II recording a grant of fields on a total Eclipse of the Sun in *Śravaṇa* (Fleet calculates that it was 25th June 754 A.D.); E.I., Vol. III, pp. 103-110. *Paithān* Plates of *Rāṣṭrakūṭa Gōvindarāja*-III recording the grant of a village after the king bathed in the *Gōdāvari* on a total Solar Eclipse in *Vaiśākha* of *Śaka* year 716 (4th May 794 A.D.); E.I., Vol. III, pp. 202-208 Sirūr Inscription of the time of *Rāṣṭrakūṭa Amōghavarṣa*-I on a Solar Eclipse in *Jyēṣṭha* of *Śaka* year 788 (expired) on a Sunday (16th June 866 A.D.); E.I., Vol. IX, pp. 98-102 grant of a village by Western *Cāḷukya* King *Vikramāditya*-I on a Solar Eclipse in *Śravaṇa* in his sixth Regnal Year (Monday, 13th July 660 A.D. according to Keilhorn); E.I., Vol. XIV, pp. 156-163 the *Naihati* grant of the *Bengal* King *Ballālasēna* of a village as a *Dakṣiṇa* on the gift of a golden horse on the occasion of a Solar Eclipse by the king's mother. For the Lunar Eclipses, the references are given in the tabular statements.

As regards *Śrāddha* it is often difficult to perform at the time of Eclipse for two reasons. Most Eclipses are partial and the time is short. Besides, there is a prohibition to eat food during an Eclipse. A *Prājāpatya* expiation is prescribed for partaking of food during an Eclipse.[161] Though the person performing *Śrāddha* in an Eclipse is deemed to reap great merit, the person partaking of *Śrāddha* dinner is liable to undergo expiation and is generally looked down upon. The *Mitākṣara* on *Yājñavalkya Smṛti,* I, 217-218 quotes the first quarter of a verse 'one should not eat during the Solar or

Lunar Eclipses'.[162] The order of the several acts on an Eclipse is: first bath in *Ganges* or other water, P*rāṇāyāma, Tarpaṇa, Japa* of *Gāyatri, Hōma* in fire with sesame with the *Vyāhṛtis* and the *Mantras* for the Planets as in *Yājñavalkya Smṛti,* I. 300-301, then *Āmaśrāddha,* gifts of gold, food, cows, and land.

An Eclipse is the best time for *Japa* and for *Dīkṣa* (initiation) and perfection in the *mantras* peculiar to various deities : One should engage in *Japa* and the like while an Eclipse of the Sun or Moon is in progress; one should not bathe nor take food during that time, but when the Sun and the Moon are free from Eclipse one should bathe and partake of food: One may engage in the *Japa* of *Gāyatri mantra* (R.V., III, 62. 10) and it is laid down that if one doesn't engage in *Japa* on an Eclipse one becomes sinful. For *Dīkṣa* as to *mantras* seven days from Eclipse are allowed. Solar Eclipse is the best time for *Dīkṣa.*[163]

The *Puṇyakāla* (the holy period) in the case of Eclipses lasts only as long as the Eclipse is visible to the eye. *Jābāli* says,[164] 'In the case of *Saṅkrānti* the *Puṇyakāla* is 16 *Kālas* on both sides thereof, but in the case of a Lunar or Solar Eclipse it lasts only so long as the Eclipse is visible'.

If a Solar Eclipse occurs on a Sunday and a Lunar one on Monday, such a conjunction was called *Cūḍāmaṇi* and it was laid down that a *Cūḍāmaṇi* Eclipse yields one crore of times as much merit as an Eclipse on other week days.[165] Some held that on the day previous to an Eclipse, one should observe a fast; but *Hēmādri* provided that a fast was to be observed on the day of the Eclipse.[166] A house holder, however, who had a son living was not to observe a fast as laid down in a text.[167]

About the partaking of food before, during and after an

Eclipse elaborate rules were laid down from comparatively early times. It was prescribed that not only was one not to eat during an Eclipse, but in the case of a Lunar Eclipse one was not to eat for three *Praharas* (9 hours or 221 *Ghaṭikas*) before the Eclipse started and for four *Praharas* before a Solar Eclipse starts, but this doesn't apply to children, old men and women.[168]

Certain astrological results were deemed to follow from Eclipses. The *Viṣṇu Dharmōttara*[169] states that if in one and the same month there is first an Eclipse of the Moon and then of the Sun, that occurrence would tend to create disputes between *Brāhmaṇas* and *Kṣatriyas,* but if the opposite was the case, then there would be prosperity. The same *Purāṇa* says that persons born on the *Nakṣatra* on which there is an Eclipse of the Sun or Moon suffer troubles unless they perform *Śānti* rites to avert the evil results.

Saṅkrānti

Saṅkrānti means, 'the passage of the Sun from one *Rāśi* (sign of the Zodiac) to the next' and hence the *Rāśi* in which the Sun enters is designated as the *Saṅkrānti* of that name.[170] When the Sun leaves the *Rāśi* called *Dhanus* and enters the *Makara Rāśi* that is called *Makara Saṅkrānti*. There are only twelve *Saṅkrāntis* even if there be an Inter-Calary month in same year. Each one of the twelve *Saṅkrāntis* is regarded as holy day. The whole Chapter 98 of *Matsya Purāṇa* deals with the completion of *Saṅkrānti Vrata.*[171] Great merit was attached to a bath in the *Ganges* on a *Saṅkrānti,* on an Eclipse and on New and Full Moon, as such a man would reach the world of *Brahma*.

The entrance of the Sun alone in a *Rāśi* is not regarded as holy time *(Puṇya Kāla),* but the entrance of all planets in a *Nakṣatra* or a *Rāśi* was deemed to be a holy time.[172] The

Kālanirṇaya specifies the holy times of the passage of the Sun and the planets as : 'In the case of the Sun, the *Puṇyakāla* is sixteen *Ghaṭikas* before and after the movement of the *Saṅkrānti;* for the Moon, it is one *Ghaṭi* and thirteen *Palas* on both sides, for Mars four *Ghaṭikas* and one *Pala,* for Mercury three *Ghaṭikas* and fourteen *Palas,* for Jupiter four *Ghaṭikas* and thirty seven *Palas,* for Venus four *Ghaṭikas* and one *Pala,* for Saturn eighty two *Ghaṭikas* and seven *Palas'.*

The twelve *Saṅkrāntis* in a year are grouped into four classes viz., two *Āyana Saṅkrāntis* (i.e., *Makara Saṅkrānti* from which *Udagayana* starts and *Karkaṭa Saṅkrānti* from which *Dakṣiṇāyana* starts), two *Viṣuva Saṅkrāntis* (i.e., *Mēṣa* and *Tula Saṅkrāntis* when the Day and Night are equal in length), the four *Saṅkrāntis* called *Ṣaḍaśīti* or *Ṣaḍaśīti-Mukha* (that is *Mithuna, Kanya, Dhanus* and *Mīna*) and *Viṣṇupadi* or *Viṣṇupada* (i.e., *Vṛṣabha, Simha, Vṛścika* and *Kumbha*).[173] The *Dēvi Purāṇa* says, '*Dēvi* is all pervading on account of the division into very small and large parts in the form of *Year, Āyana, Season, Month, Fortnight, Day* and the like. *Dēvi* bestows rewards in accordance with the divisions into meritorious and sinful actions, even one act done at that time of *Saṅkrānti* yields results multiplied crores of times. From *Dharma* (righteous acts) increase life, kingdom, progeny, happiness and the like, from *Adharma* (unrighteous conduct) spring diseases, sorrows and the like; whatever is donated and whatever Japa is performed near *Viṣuva (Mēṣa* and *Tula) Saṅkrāntis* becomes inexhaustible; the same holds good about *Viṣṇupada* and *Ṣaḍaśīti-Mukha*.[174]

It is laid down that in *Saṅkrāntis,* the *Puṇyakāla* is seven folds,[175] viz., three, four, five, seven, eight, nine or twelve *Ghaṭikas;* during those periods real merit is accumulated. If one can not perform the prescribed rites within these limits, thirty *Ghaṭikas* are permitted as the highest limit, for all *Saṅkrāntis* except *Ṣaḍaśīti* (in which the maximum *Puṇyakāla* is 60 *Ghaṭikas*) and *Viṣṇupadi* (where it is 16 *Ghaṭikas* before and after).

Each of these *Saṅkrāntis* in a year Punyakāla is seven folds. Accor-dingly, it occurs on a particular weekday or in conjunction with certain classes of *Nakṣatra*. The seven names[176] are : *Manda, Mandākini, Dhvāṅkṣi, Ghōra, Mahōdari, Rākṣasi,* and *Miśrita.* A *Saṅkrānti,* such as *Mēṣa* or *Karka* or *Makara* is called *Ghōra* if it occurs on Sunday, *Dhvāṅkṣi* on Monday, *Mahōdari* on Tuesday, *Mandākini* on Wednesday, *Manda* on Thursday, *Miśrita* on Friday, and *Rākṣasi* on Saturday. Again, a *Saṅkrānti* like *Mēṣa* or *Karka* and so on is called respectively *Manda, Mandākini, Dhvāṅkṣi, Ghōra, Mahōdari, Rākṣasi, Miśrita* accordingly as it commences respectively on *Nakṣatras* called *Dhṛva, Mṛdu, Kṣipra, Ugra, Cara, Krūra.* It is laid down that *Manda Saṅkrānti* is beneficial to *Brāhmaṇas, Mandākini* to *Kṣatriyas, Dhvāṅkṣi* to *Vaiśyā, Ghōra* to *Śūdras, Mahōdari* to thieves, *Rākṣasi* brings success to vendors of liquor and *Miśrita* increases the profit of the avocations of *Caṇḍālas, Pukkasas,* of those whose business is of a terrible nature and of all craftsmen.

A bath is to be taken and gifts are to be made during that time in order to secure the full merit thereof. The general rule is that bath is not to be taken and a gift is not to be made at night. The *Bhaviṣya Purāṇa* provides, 'one should not take a bath at night and particularly should not make a gift at night, but a bath and a gift may be made at night on proper occasions. A gift at night is commended in an Eclipse, marriage, *Saṅkrānti,* pilgrimage, on death, on the birth of a child and at hearing *Itihāsa (Mahābhārata)'.* There are special recommendations about gifts in various texts-gift of ram on

Mēṣa; of cows on *Vṛṣabha;* of clothes, food and drinks on *Mithuna;* of *Ghṛtadhēnu* on *Karkaṭa;* of a conveyance with some gold on *Simha*; of clothes and cows on *Kanya;* of different kinds of grains and seeds on *Tula;* on *Vṛścika* of clothes and houses; on *Dhanus* of clothes and conveyance; on *Makara* of wood and fire; on *Kumbha* of cows, water and grass; on *Mīna* of fresh flowers. Special gifts are required to be made on *Makara Saṅkrānti*. The *Skandha Purāṇa* says, 'He who donates *Tilādān* on *Uttarāyaṇa* (i.e., *Makara Saṅkrānti*) secures all objects and obtains the highest happiness.[178] Sesame should be given to *Brāhmaṇas* on *Uttarāyaṇa Saṅkrānti*. A Person should also observe a fast for three days or one day in honour of *Makara Saṅkrānti*'. *Āpastambha* says, 'A man who after a bath fasts the whole day on *Āyana, Viṣuva* and on an eclipse of the Sun or Moon becomes free from all sins'. Epigraphical evidences are enlisted in the tabular statements of this Chapter.

In modern times *Makara Saṅkrānti* is more a social festival than a religious one. In South India, there is festival called *Poṅgal,* which corresponds with *Makara Saṅkrānti* observed in North India and Western India.

REFERENCES

1. संवत्सरीणं पय उस्त्रियायास्तस्य माशीद्यातुधानो नृचक्षः।
पीयूषमग्ने यतमस्तितृप्सात् तं प्रत्यञ्चमर्चिषा विध्या मर्मन्॥

ऋग्वेद, १०.८७.१७.

Saṃvatsarīṇaṃ Paya Usriyāyāstasya Māśīdyātudhānō Nṛcakṣaḥ. Pīyūṣamagne Yatamastitṛpsāt Taṃ Pratyañcamarciṣā Vidhyā Marman.

Ṛgvēda, X. 87. 17.

2. धनुस्तद्वातो अनु वातु ते तस्मै ते रूद्र संवत्सरेण नमस्करोमि यत् ते रूद्र दक्षिणा धनुस्तद्वातो अनु वातु ते तस्मै ते रूद्र परिवत्सरेण नमस्करोमि यत् ते रूद्र पश्चाद्धनु-स्तद्वातो अनु वातु ते तस्मै ते रूद्रेदावत्सरेण नमस्करोमि यत् ते

रूद्रोत्तराद्धनुस्तद्वातो अनु वातु ते तस्मै ते रूद्रेदुवत्सरेण नमस्करोमि यत् ते रूद्रोपरिद्धनुस्तद्वातो अनु वातु ते तस्मै ते रूद्र वत्सरेण नमस्करोमि रूद्रो वा एष यदग्निः स यथा व्याघ्रः ऋद्धस्तिष्ठत्येवं वा एष एतर्हि संचितमेतैरूपतिष्ठते नमस्कारै रेवैनं शमयति॥

तैत्तिरीय संहिता, ५. ५. ७. १-३.

Dhanustadvātō Anu Vātu Tē Tasmai Tē Rudra Saṃvatsarēṇa Namaskarōmi Yat Tē Rudra Dakṣiṇā Dhanustadvātō Anu Vātu Tē Tasmai Tē Rudra Parivatsarēṇa Namaskarōmi Yat Tē Rudra Puścāddhanustadvāto Anu Vātu Te Tasmai Te Rudredavatsareṇa Namaskarōmi Yat Tē Rudrōttarāddhanustadvātō Anu Vātu Tē Tasmai Tē Rudrēduvatsarēṇa Namaskarōmi Yat Tē Rudrōpari Dhanustadvātō Anu Vātu Tē Tasmai Te Rudra Vatsarēṇa Namaskarōmi Rudrō Vā Ēṣa Yadagniḥ Sa Yathā Vyāghraḥ Ṛuddhastiṣṭ hatyēvaṃ Vā Ēṣa Ētarhi Sañcitamētairupa Tiṣṭhatē Namaskārai Rēvainaṃ Śamayati.

Taittirīya Saṃhitā, V. 5. 7. 1-3.

3. अग्निर्वाव संवत्सरः। आदित्यः परिवत्सरः। चन्द्रमा इदावत्सरः।
वायुः पुनरनुवत्सरः यद्वैश्वदेवेन यजते। अग्निमेव तत्संवत्सरमाप्नोति।
तस्माद्वैश्वदेवेन यजमानः संवत्सरीणां स्वस्तिमाशास्त इत्याशासीत। यद्वरूण-प्रधासैर्यजते।
आदित्यमेव तत्परिवत्सरमाप्नोति। यत्साकमेधैर्यजते चंद्रमसमेव तदिदावत्सर-माप्नोति।
...यत्पितृपक्षेन यजते देवानेव तदन्ववस्यति।
अथवा अस्य वायुश्वानुवत्सर श्वाप्रीतावुच्छिष्यते।
यच्छुनासीरीयेण यजते वायुमेव तवनुवत्सरमाप्नोति।

तैत्तिरीय ब्राह्मणम्, १. ४. १०. १-३.

Agnirvāva Saṃvatsaraḥ। Ādityaḥ Parivatsaraḥ। Candramā।
Idāvatsaraḥ। Vāyuḥ Punaranuvatsaraḥ Yadvaiśvadēvēna Yajatē।
Agnimēva Tatsaṃvatsaramāpnōti।
Tasmādvaiśvadēvēna Yajamānaḥ Saṃvatsarīṇāṃ Svastimāśāsta Ityāśāsīta। Yadvaruṇapeadhāsairyajatē। Ādityamēva Tatparivatsara-māpnōti।
Yatsākamedhairyajatē Candramasamēva Tadidāvatsaramāpnōti।
...Yatpitṛpakṣēna Yajatē Dēvānēva Tadanvavasyati।
Athavā Asya Vāyuśvānuvatsara Śvāprītāvucchiṣyatē।
Yacchunāsīriyēṇa Yajatē Vāyumēva Tavanuvatsaramāpnōti।

Taittirīya Brāhmaṇa, I. 4. 10. 1-3.

4. पंचसंवत्सरोयुगमिति।
...एवमर्धतृतीयानामब्दानामधि मासकम्
ग्रीष्मे जनयतः पूर्व पंचाब्दान्ते च पश्चिमम्॥
अर्थशास्त्रम्, II Chapter 20 (देशकालमान) p. 109.

Pañca samvatsarōyugamiti l
...Ēvamardhatṛtīyānāmabdānmadhi Māsakam l
Grīṣmē Janayataḥ Pūrva Pañcābdāntē Ca Paścimam ll
Arthaśāstra. II. Chapter 20 (Dēśakālamāna) p. 109.

5. क्षण लवा मुहूर्ताश्च दिवारात्रिस्तथैव च।
अर्धमासाश्च ऋतवः षट् च भारतः।
संवत्सराः पञ्चयुगमहोरात्रश्चतुर्विधः।
महाभारतम्, सभापर्वम्, २.३७-३८

Kṣaṇa Lavā Muhūrtāśca Divārātristathaiva ca l
Ardhamāsāśca Ṛtavaḥ Ṣaṭ Ca Bhārataḥ ll
Saṃvatsarāḥ Pañcayugamahōrātraścaturvidhaḥ l
Mahābhārata, Sabhāparvan II. 37-38.

6. द्वादशारं नहि तज्जराय वर्वर्ति चक्रं परि द्यामृतस्य।
आ पुत्रा अग्नेमिथुनासो अत्र सप्त शतानि विम्शतिश्च तस्थुः॥
ऋग्वेद, १.१६४.११.

Dvādaśāraṃ Nahi Tajjarāya Varvarti Cakraṃ Pari Dyāmṛtasya l
Ā Putrā Agnēmithunāsō Atra Sapta Śatāni Vimśatiśca Tasthuḥ ll
Ṛgvēda. I. 164.11.

पञ्चपादं पितरं द्वादशाकृतिं दिव आहुः परे अर्धे पुरीषिणम्।
अथेमे अन्य उपरे विचक्षणं सप्तचक्रे षळर आहुरर्पितम्॥१२॥
Pañcapādaṃ Pitaraṃ Dvādaśākṛtiṃ Diva Āhuḥ Parē Ardhē Purīṣham l
Athēmē Anya Uparē Vicakṣaṇaṃ Saptacakrē Ṣaḷara Āhurarpitam ll12ll
पञ्चारे चक्रे परिवर्तमाने तस्मिन्ना तस्थुर्भुवनानि विश्वा।
तस्य नाक्षस्तप्यते भूरिभारः सनादेव न शीर्यते सनाभिः॥१३॥
Pañcārē Cakrē Parivartamānē Tasminnā Tasthurbhuvanāni Viśvā l
Tasya Nākṣastapyatē Bhūribhāraḥ Sanādēva Na Śīryatē Sanābhiḥ ll13ll
द्वादश प्रधयश्चक्रमेकं त्रीणि नभ्यानि क उ तच्चिकेत।
तस्मिन् त्साकं त्रिशता न शङ्कवोर्पिताः षष्टिर्न चलाचलासः॥४८॥

Dvādaśa Pradhayaścakramēkaṃ Trīṇi Nabhyāni Ka U Taccikēta I Tasmin Tsākaṃ Triśatā Na Śaṅkavōrpitāḥ Ṣaṣṭirna Calācalāsaḥ II48II

7. यस्मान्मासा निर्मितास्त्रिंशदराः संवत्सरो यस्मान्निर्मितो द्वादशारः।
अहोरात्रा यं परियन्तो नापुस्तेनौदनिनाति तराणि मृत्युम्॥

अथर्ववेद, ५.३५. ४.

Yasmānmāsā Nirmitāstriṃśadarāḥ Saṃvatsarō Yasmānnirmitō Dvādaśāraḥ I

Ahōrātrā Yaṃ Pariyantō Nāpustēnaudaninàti Tarāṇi Mṛtyum II

Atharvavēda, V. 35. 4.

8. त्रीणि च शतानि षष्टिश्चानूच्यानि यज्ञकामस्य।
त्रीणि च वै शतानि षष्टिश्च शंवत्सरस्याहानि॥
तावान् संवत्सरः। संवत्सरः प्रजापतिः। प्रजापतिर्यज्ञः।
उपैनं यज्ञो नमति यस्यैवं विद्वान् त्रीणि च शतानि षष्टिम् चान्वाह॥
सप्त च शतानि विंशतिश्चानूच्यानि प्रजापशुकामस्य।
सप्त च वै शतानि विंशतिश्च संवत्सरस्याहोरात्राः॥

ऐतरेय ब्राह्मणम्, ७.७.

Trīṇi Ca Śatāni Ṣaṣṭiścānūcyāni Yajñakāmasya I

Trīṇi Ca Vai Śatāni Ṣaṣṭiśca Śaṃvatsarasyāhāni II

Tāvān Saṃvatsaraḥ I Saṃvatsaraḥ Prajāpatiḥ I Prajāpatiryajñaḥ I Upainaṃ Yajñō Namati Yasyaivaṃ Vidvān Trīṇi Ca Śatāni Ṣaṣṭim Cānvāha II

Sapta Ca śatāni Viṃśatiścānūcyāni Prajāpaśukāmasya I

Sapta Ca Vai Śatāni Viṃśatiśca Saṃvatsarasyāhōrātrāḥ II

Aitarēya Brāhmaṇa, VII. 7.

तदाहुः कथमस्यैतच्छत रूद्रियं संवत्सरं अग्निमाप्नोति कथं संवत्सरेणाग्निना संपद्यत इति षष्टिश्च ह वै त्रीणि च शतान्येतच्छतरूद्रियमथ त्रिंशदथ पञ्चत्रिंशत्। ततो यानि षष्टिश्च त्रीणि च शतानि तावन्ति संवत्सरस्याहानि तत्संत्सरस्याहान्याप्नोति। अथ यानि त्रिंशत् त्रिंशन्मासस्य रात्रयस्तन्मासस्य रात्रीराप्नोति तदुभयानि संवत्सरस्याहोरात्राण्याप्नोति अथ यानि पञ्चत्रिंशत्स त्रयोदशो मासः स आत्मा त्रिंशदात्मा प्रतिष्ठा द्वे प्राणा द्वे शिर एव पञ्चत्रिंशमेतावान्वै संवत्सर एवमु हस्यैतच्छत रूद्रियं संवत्सरं अग्निमाप्नोत्येवं संवत्सरेण अग्निना संपद्यतएतवत्य उ वै शाण्डिलेग्नौ मध्यतो यजुष्मत्य इष्टका उपधीयन्तेग्नयो हैते पृतग्यदेता इष्टका एवसु हास्यैतेग्नयः पृथक्छतरूद्रियेणाभिहुता भवन्ति।

शतपथ ब्राह्मणम्, ९.१.१.४३.

Tadāhuḥ Kathamasyaitacchata Rudriyaṃ Saṃvatsaraṃ Agnimāp-

nōti Katham Saṃvatsarēṇāgninā Sampadyata iti Ṣaṣṭiśca Ha Vai Trīṇi Ca Śatānyētacchatarudriyamatha Triṃśadatha Pañcatriṃśat। Tatō Yāni Ṣaṣṭiśca Trīṇi Ca Śatāni Tāvanti Saṃvatsarasyāhāni Tatsaṃvatsarasyāhānyāpnōti। Atha Yāni Triṃśat Triṃśanmāsasya Rātrayastanmāsasya Rātrīrāpnōti Tadubhayāni Saṃvatsarasyāhōrāṇyāpnōti Atha Yāni Pañcatriṃśatsa Trayōdaśō Māsaḥ Sa Ātmā Triṃśadātmā Pratiṣṭhā Dvē Prāṇā Dvē Śira Ēva Pañcatriṃśamētāvānvai Samvatsara Ēvamu Hasyaitaccata Rudriyam Samvatsaram Agnimāpnōtyēvam Saṃvatsarēṇa Agninā Sampadyataētavatya U Vai Śāṇḍilegnau Madhyatō Yajuṣmatya Iṣṭakā Upadhīyantēgnayō Haitē Pṛtagyadētā Iṣṭakā Ēvasu Hāsyaitēgnayaḥ Pṛthakchatarudriyēṇābhihutā Bhavanti।

Śatapatha Brāhmaṇa, IX. 1. 1. 43.

9. अग्नेर्वै दीक्षया देवा विरजमाप्नुवन् तिस्रो रात्रीर्दीक्षितः स्यत् त्रिपदा विराड् विराजमाप्नोति षड्रात्रीर्दीक्षितः स्यत् षड् वा ऋतवः संवत्सरः संवत्सरो विराड् विराजमाप्नोति दश रात्रीर्दीक्षितः स्याद्दशाक्षरा विराड् विराजमाप्नोति द्वादश रात्रीर्दीक्षितः स्याद् द्वादश मासाः संवत्सरः संवत्सरो विराड् विराजमाप्नोति त्रयोदश रात्रीर्दीक्षितः स्यात् त्रयोदश॥

तैत्तिरीय संहिता, ५. ६. ७. १.

Agnērvai Dīkṣayā Dēvā Virajamāpnuvan Tisrō Rātrīrdīkṣitaḥ Syat Tripadā Virāḍ Virājamāpnōti Ṣaḍrātrīrdīkṣitaḥ Syat Ṣaḍ Vā Ṛtavaḥ Saṃvatsaraḥ Saṃvatsarō Virāḍ Virājamāpnōti Daśa Rātrīrdīkṣitaḥ Syāddaśākṣarā Virāḍ Virājamāpnōti Dvādaśa Rātrīrdīkṣitaḥ Syād Dvadaśa Māsāḥ Saṃvatsaraḥ Saṃvatsarō Virāḍ Virājamāpnōti Trayōdaśa Rātrīrdīkṣitaḥ Syāt Trayōdaśa।

Taittirīya Saṃhitā, V. 6. 7. 1.

मासाः सवत्सरः संवत्सरो विराड् विराजमाप्नोति पञ्चदश रात्रीर्दीक्षितः स्यात् पञ्चदश वा अर्धमासस्य रात्रयोर्धमासशः संवत्सर अप्यते संवत्सरो विराड् विराजमाप्नोति सप्तदश रात्रीर्दीक्षितः स्याद् द्वादश मासाः पञ्चर्तवः स संवत्सरः संवत्सरो विराड् विराजमाप्नोति चतुर्विंशतिंरात्रीर्दीक्षितः स्यच्चतुर्विंशतिरर्धमासाः संवत्सरः संवत्सरो विराड् विराजमाप्नोति त्रिंशतँ रात्रीर्दीक्षतः स्यात्॥२॥

Māsāḥ Saṃvatsaraḥ Saṃvatsarō Virāḍ Virājamāpnōti Pañcadaśa Rātrīrdīkṣitaḥ Syāt Pañcadaśa Vā Ardhamāsasya Rātrayōrdhamāsaśaḥ Saṃvatsara Apyatē Saṃvatsarō Virāḍ Virājamāpnōti Saptadaśa Rātrīrdīkṣitaḥ Syād Dvādaśa Māsāḥ Pañcartavaḥ sa Samvatsaraḥ Samvatsarō virāḍ virājamāpnōtī Caturvimśatimrātrīrdīkṣtiḥ Syaccaturvimśatirardhamāsāḥ Saṃvatsaraḥ Saṃvatsarō Virāḍ Virājamāpnōti Trimśatam Rātrīrdīkṣitaḥ Syāt।।2।।

10. Kauṣītakī Brāhmaṇa, XIX. 2.
11. त्रयोदशं वा एतं मासमाप्नोति यच्छुनासीर्येण यजत एतावान्वै सं वत्सरो यदेष त्रयोदशो मासस्तदत्रैव सर्वः संवत्सर आप्तो भवति॥

कौषीतकी ब्राह्मणम्, ५. ८.

Trayōdaśaṃ Vā Ētaṃ Māsamāpnōti Yacchunāsīryēṇa Yajata Ētāvānvai Saṃ Vatsarō Yadēṣa Trayōdaśō Māsastadatraiva Sarvaḥ Saṃvatsara Āptō Bhavati//

Kauṣītakī Brāhmaṇa, V. 8.

12. युधिष्ठिरो विक्रमशालिवाहनौ नराधिनाथौ विजयाभिनंदनः इमे नु नागार्जुन मेदिनीविभुर्बलि क्रमात् षट्।

शक कारकाः कालौ॥

Yudhiṣṭirō Vikramaśālivāhanau Narādhināthau Vijayābhinandanaḥ lmē Nu Nāgārjuna Mēdinīvibhurbali Kramāt Ṣaṭ/

Śaka Kārakāḥ Kālau//

Fleet in J.R.A.S. for 1911, p. 694.

13. Corpus of Inscriptionum Indicarum (C.I.I.). Vol., I. pp. 147. 151.
14. राजवर्षं मासः पक्षो दिवसश्च व्युष्टं वर्षाहेमंतग्रीष्माणानि तृतीयसप्तमा दिवसेनाः। पक्षाः शेषाः पूर्णाः पृथगधिमासक इति कालः॥

अर्थशास्त्रम्, २. ६. पृ०. ६०.

Rājavarṣaṃ Māsaḥ Pakṣō Divasaśca Vyuṣṭaṃ
Varṣāhēmantagrīṣmāṇāni Tṛtīyasaptamā Divasēnāḥ/
Pakśāḥ Śēṣāḥ Pūrṇāḥ Pṛthagadhimāsaka Iti Kālaḥ//

Arthaśāstra, II.6. p. 60.

15. Smṛti Candrika on Śrāddha, p. 377.
16. Epigraphia Indica (E.I.). Vol. XXIX. Part V. pp. 116-117.
17. मागशुक्लं समारभ्य चन्द्रार्कौ वासवर्क्षगौ।
जीवयुक्तौ यदा स्यातां षष्ट्यब्दादिस्तदा स्मृतः।

विष्णुधर्मोत्तर पुराणम्, १, ८२. ८.

Māgaśuklaṃ Samārabhya Candrārkau Vāsavarkṣagau/
Jīvayuktau Yadā Syātāṃ Ṣaṣṭyabdādistadā Smṛtaḥ//

Viṣṇudharmōttara Purāṇa, I, 82.8.

18. Indian Antiquary (I.A.). XIX. p. 1. ff at, p. 18.
19. E.I., Vol. VI. p. 209.
20. आद्यं धनिष्ठांशमभिप्रपन्नो माघे यदा यात्युदयं सुरेज्यः।
षष्ठ्यब्दपूर्वः प्रभवः स नाम्ना प्रपद्यते भूतहितस्तदाब्दः॥

बृहत्संहिता, ८. २७.

Ādyaṃ Dhaniṣṭhāṃśamabhiprapannō Māghē Yadā yātyudayaṃ Surējyaḥ/

Ṣaṣṭyabdapūrvaḥ Prabhavaḥ Sa Nāmnā Prapadyatē Bhūtahitasta-dābdaḥ//

Bṛhatsaṃhitā. 8.27.

क्वचित्ववृष्टिः पवनाग्निकोपः सन्तीतयः श्लेष्मकृताश्च रोगाः।
संवत्सरेस्मिन्प्रभवे प्रवृत्ते न दुःखमाप्नोति जनस्तथापि॥२८॥

Kvacitvavṛṣṭiḥ Pavanāgnikōpaḥ Santītayaḥ Ślēṣmakṛtāśca Rōgāḥ/
Saṃvatsarēsminprabhavē Pravṛttē Na Duḥkhamāpnōti Janasta-thāpi//28//

तस्माद् द्वितीयो विभवः प्रदिष्टः शुक्लस्तृतीयः परतः प्रमोदः।
प्रजापतिश्चेति यतोत्तराणि शस्तानि वर्षाणि फलान्यथैषाम्॥२९॥

Tasmād Dvitīyō Vibhavaḥ Pradiṣṭaḥ Śuklastṛtīyaḥ Parataḥ Pramō-daḥ/
Prajāpatiścēti Yatōttarāṇi Śastāni Varṣāṇi Phalānyathaiśām//29//

निष्पन्नशालीक्षुयवादिसस्यां भयैर्विमुक्तामुपशान्तवराम्।
संहृष्टलोकां कलिदोषमुक्तां क्षत्रं तदा शास्ति च भूतधात्रीम्॥३०॥

Niṣpannaśālīkṣuyavādisasyāṃ Bhayairvimuktāmupaśāntavarām/
Saṃhṛṣṭalōkāṃ Kalidōṣamuktāṃ Kṣatraṃ Tadā Śāsti Ca Bhūta-dhātrīm//30//

आध्योङ्गिराः श्रीमुखभावसाह्वौ युवा सुधातेति युगे द्वितीये।
वर्षाणि पञ्चैव यथाक्रमेण त्रीण्यत्र शस्तानि समे परे द्वे॥३१॥

Ādhyōṅgirāḥ Śrīmukhabhāvasāhvau Yuvā Sudhātēti Yugē Dvitīyē/
Varṣāṇi Pañcaiva Yathākramēṇa Trīṇyatra Śastāni Samē Parē Dvē//31//

त्रिष्वाद्यवर्षेषु निकामवर्षो देवो निरातङ्कभयश्च लोकः।
अब्दद्वयेन्त्येपि समा सुवृष्टिः किन्त्वत्र रोगाः समरागमश्च॥३२॥

Triṣvādyavarṣēṣu Nikāmavarṣō Dēvō Nirātaṅkabhayaśca Lōkaḥ/
Abdadvayēntyēpi Samā Suvṛṣṭiḥ Kintvatra Rōgāḥ Samarāga-maśca//32//

शाक्रे युगे पूर्वमथेश्वराख्यं वर्षं द्वितीयं बहुधान्यमाहुः।
प्रमाथिनं विक्रममप्यथान्यद्वृषं च विन्द्याद्गुरुचारयोगात्॥३३॥

Śākrē Yugē Pūrvamathēśvarākhyaṃ Varṣa Dvitīyaṃ Bahudhānya-māhuḥ/
Pramāthinaṃ Vikramamapyathānyadvaṣaṃ Ca Vindyādgurucāra Yōgāt//33//

आद्यं द्वितीयं च शुभे तु वर्षे कृतानुकारं कुरूतः प्रजानाम्।
पापः प्रमाथी वृषविक्रमौतु सुभिक्षदौ रोगभयप्रदौ च॥३४॥

Ādyaṃ Dvitīyaṃ Ca Śubhē Tu Varṣē Kṛtānukāraṃ Kurutaḥ Prajānām/

Pāpaḥ Pramāthī Vṛṣavikramautu Subhikṣadau Rōgabhayapradau Ca//34//

श्रेष्ठं चतुर्थस्य युगस्य पूर्व यच्चित्रभानुं कथयन्ति वर्षम्।
मध्यं द्वितीयं तु शुभानुसञ्ज्ञं रोगप्रदं मृत्युकरं न तच्च॥३५॥

Śrēṣṭhaṃ Caturthasya Yugasya Pūrva Yaccitrabhānuṃ Kathayanti Varṣam/
Madhyaṃ Dvitīyantu Śubhānusañjñaṃ Rōgapradaṃ Mṛtyukaraṃ Na Tacca//35//

तारणं तदनु भूरिवारिदं सस्यवृद्धिमुदितं च पार्थिवम्।
पञ्चमं व्ययमुशन्ति शोभनं मन्मथप्रबलमुत्सवाकुलम्॥३६॥

Tāraṇaṃ Tadanu Bhūrivāridaṃ Sasyavṛddhimuditaṃ Ca Pārthivam/
Pañcamaṃ Vyayamuśanti Śōbhanaṃ Manmathaprabalamutsavākulam//36//

त्वाष्ट्रे युगे सर्वजिदाद्य उक्तः संवत्सरोन्यः खलु सर्वधारी।
तस्माद्विरोधी विकृतः खरश्च शस्तो द्वितीयोत्र भयाय शेषाः॥३७॥

Tvāṣṭrē Yugē Sarvajidādya Uktaḥ Saṃvatsarōnyaḥ Khalu Sarvadhārī/
Tasmādvirōdhi Vikṛtaḥ Kharaśca Śastō Dvitīyōtra Bhayāya Śēṣāḥ//37/

नन्दनोथ विजयो जयस्तथा मन्मथोस्य परतश्च दुर्मुखः।
कान्तमत्र युग आदितस्त्रयं मन्मथः समफलोधमोपरः॥३८॥

Nandanōtha Vijayō Jayastathā Manmathōsya Parataśca Durmukhaḥ/
Kāntamatra Yuga Āditastrayaṃ Manmathaḥ Samaphalōdhamōparaḥ//38//

हेमलम्ब इति सप्तमे युगे स्याद्विलम्बि परतो विकारि च।
शर्वरीति तदनु प्लवः स्मृतो वत्सरो गुरुवशेन पञ्चमः॥३९॥

Hēmalamba Iti Saptamē Yugē Syādvilambi Paratō Vikāri Ca/
Śarvarīti Tadanu Plavaḥ Smṛtō Vatsarō Guruvaśēna Pañcamaḥ//39//

ईतिप्राया प्रचुरपवना वृष्टिरब्दे तु पूर्वे
मन्दं सस्यं न बहुसलिलं वत्सरेतो द्वितीये।
अत्युद्वेगः प्रचुरसलिलः स्यत्तृतीयशचतुर्थो
दुर्भिक्षाय प्लव इति ततः शोभनो भूरितोयः॥४०॥

Ītiprāyā Pracurapavanā Vṛṣtirabdē Tu Pūrvē
Mandaṃ Sasyaṃ Na Bahusalilaṃ Vatsarētō Dvitīyē/
Atyudvēgaḥ Pracurasalilaḥ Syatṛtīyaśacaturthō
Durbhikṣāya Plava Iti Tataḥ Śōbhanō Bhūritōyaḥ//40//

वैश्वे युगे शोभकृदित्यथाद्यः संवत्सरोतः शुभकृद्द्वितीयः।
क्रोधी तृतीयः परतः क्रमेण विश्वावसुश्चेति पराभवश्च॥४१॥

Vaiśvē Yugēśōbhakṛdityathādyaḥ Saṃvatsarōtaḥ Śubhakṛddvitīyaḥ I
Krōdhī Tṛtīyaḥ Parataḥ Kramēṇa Viśvāvasuścēti Parābhavaśca II41 II

पूर्वापरौ प्रीतिकरौ प्रजानामेषां तृतीयो बहुदोषदोब्दः।
अन्त्यौ समौ किन्तु पराभवेग्निः शस्त्रामयार्तिर्द्विजगोभयं च॥४२॥

Pūrvāparau Prītikarau Prajānāmēṣāṃ Tṛtīyō Bahudōṣadōbdaḥ I
Antyau Samau Kintu Parābhavēgniḥ Śastrāmayārtirdvijagōbhayaṃ Ca II42 II

आद्यः प्लवङ्गो नवमे युगेब्दः स्यात्कीलकोन्यः परतश्च सौम्यः।
साधारणो रोधकृदित्यथाब्दः शुभप्रदौ कीलकसौम्यसञ्ज्ञौ॥४३॥

Ādyaḥ Plavaṅgō Navamē Yugēbdaḥ Syātkīlakōnyaḥ Parataśca Saumyaḥ I
Sādhāraṇō Rōdhakṛdityathābdaḥ Śubhapradau Kīlakasaumyasañjñau II43 II

कष्टः प्लवङ्गो बहुशः प्रजानां साधारणेल्पं जलमीतयश्च।
यः पञ्चमो रोधकृदित्यथाब्दश्चित्रं जलं तत्र च सस्यसम्पत्॥४४॥

Kaṣṭaḥ Plavaṅgō Bahuśaḥ Prajānāṃ Sādhāraṇēlpaṃ Jalamītayaśca I
Yaḥ Pañcamō Rōdhakṛdityathābdaścitraṃ Jalaṃ Tatra Ca Sasyasampat II44 II

इन्द्राग्निदैवं दशमं युगं यत्तत्राद्यवर्षं परिधावि संज्ञम्।
प्रमाद्यथानन्दमतः परं यत्स्याद्राक्षसं चानलसञ्ज्ञितं च॥४५॥

Indrāgnidaivaṃ Daśamaṃ Yugaṃ Yattatrādyavarṣaṃ Paridhāvi Sañjñam I
Pramādyathānandamataḥ Paraṃ Yatsyādrākṣasaṃ Cānalasañjñitaṃ Ca II45 II

परिधाविनि मध्यदेशनाशो नृपहानिर्जलमल्पमग्निकोपः।
अलसस्तु जनः प्रमादिसञ्ज्ञे डमरं रक्तकपुष्पबीजनाशः॥४६॥

Paridhāvini Madhyadēśanāśō Nṛpahānirjalamalpamagnikōpaḥ I
Alasastu Janaḥ Pramādisañjñē Ḍamaraṃ Raktakapuṣpabījanāśaḥ II46 II

तत्परः सकललोकनन्दनो राक्षसः क्षयकरोनलस्तथा।
ग्रीष्मधान्यजननोत्र राक्षसो वह्निकोपमरकप्रदोनलः॥४७॥

Tatparaḥ Sakalalōkanandanō Rākṣasaḥ Kṣayakarōnalastathā I
Grīṣmadhānyajananōtra Rākṣasō Vahnikōpamarakapradōnalaḥ II47 II

एकादशे पिङ्गलकालयुक्तसिद्धार्थरौद्राः खलु दुर्मतिश्च।
आद्ये तु वृष्टिर्महती सचौरा श्वासो हनू कम्पयुतश्च कासः।।४८।।
Ēkādaśē piṅgalakālayuktasiddhārtharaudrāḥ Khalu Durmatiśca।
Ādyē Tu Vṛṣṭirmahatī Sacaurā Śvāsō Hanū Kampayutaśca Kāsaḥ।।48।।
यत्कालयुक्तं तदनेकदोषं सिद्धार्थसंज्ञे बहवो गुणाश्च।
रौद्रोतिरौद्रः क्षयकृत्प्रदिष्टो यो दुर्मतिर्मध्यमवृष्टिकृत्सः।।४९।।
Yatkālayuktaṃ Tadanēkadōṣaṃ Siddhārthasañjñē Bahavō Guṇāśca।
Raudrōtiraudraḥ Kṣayakṛtpradiṣṭō Yō Durmatimadhyamavṛṣikṛt-saḥ।।49।।
भाग्ये युगे दुन्दुभिसञ्ज्ञमाद्यं सस्यस्य वृद्धिं महतीं करोति।
उद्गारिसञ्ज्ञं तदनु क्षयाय नरेश्वराणां विषमा च वृष्टिः।।५०।।
Bhāgyē Yugē Dundubhisañjñamādyaṃ Sasyasya Vṛddhiṃ Mahatīṃ Karōti।
Udgārisañjñaṃ Tadanu Kṣayāya Narēśvarāṇāṃ Viṣamā Ca Vṛṣṭiḥ।।50।।
रक्ताक्षमब्दं कथितं तृतीयं यस्मिन्भयं दंष्ट्रिकृतं गदाश्च।
क्रोधं बहुक्रोधकरं चतुर्थं राष्ट्राणि शून्यीकुरूते विरोधैः।।५१।।
Raktākṣamabdaṃ Kathitaṃ Tṛtīyaṃ Yasminbhayaṃ Daṃṣṭrikṛtaṃ Gadāśca।
Krōdhaṃ Bahukrōdhakaraṃ Caturthaṃ Rāṣṭrāṇi Śūnyīkurutē Virōdhaiḥ।।51।।
क्षयमिति युगस्यान्त्यस्यान्त्यं बहुक्षयकारकं
जनयति भयं तद्विप्राणां कृषीवलवृद्धिदम्।
उपचयकरं विट्शूद्राणां परस्वहृतां तथा
कथितमखिलं षष्ट्यब्दे यत्तदत्र समासतः।।५२।।
Kṣayamiti Yugasyāntyasyāntyaṃ Bahukṣayakārakaṃ
Janayati Bhayaṃ Tadviprāṇāṃ Kṛṣīvalavṛddhidam।
Upacayakaraṃ Viṭśūdrāṇāṃ Parasvahṛtāṃ Tathā
Kathitamakhilaṃ Ṣaṣṭyabdē Yattadatra Samāsataḥ।।52।।

21. Agni Purāṇa, Chapter 139.
22. Viṣṇudharmōttara Purāṇa, I. 82. 9 ff.
23. Jyōtistattva. pp. 692-697.
24. क्वचित्ववृष्टिः पवनाग्निकोपः सन्तीतयः श्लेष्मकृताश्च रोगाः।
संवत्सरेस्मिन्प्रभवे प्रवृत्ते न दुःखमाप्नोति जनस्तथापि।।
बृहत्संहिता, ८. २८.

Kvacitvavṛṣṭiḥ Pavanāgnikōpaḥ Santītayaḥ Ślēṣmakṛtāśca Rōgāḥ। Saṃvatsarēsminprabhavē Pravṛttē Na Duḥkhamāpnōti Janastathāpi॥

Bṛhatsaṃhitā, 8. 28.

तस्माद् द्वितीयो विभवः प्रदिष्टः शुक्लस्तृतीयः परतः प्रमोदः।
प्रजापतिश्चेति यतोत्तराणि शस्तानि वर्षाणि फलान्यथैषाम्॥२९॥

Tasmād Dvitīyō Vibhavaḥ Pradiṣṭaḥ Śuklastṛtīyaḥ Parataḥ Pramōdaḥ।

Prajāpatiścēti Yatōttarāṇi Śastāni Varṣāṇi Phalānyathaiṣām॥29॥

25. आयन्नापोयनमिच्छमानाः

ऋग्वेद, ३. ३३. ७.

Āyannāpōyanamicchamānāḥ Ṛgvēda, III. 33. 7.

26. सा यत्रोदङ्ङावर्तते।
देवेषु तर्हि भवति देवांस्तर्हाभिगोपायत्यथ यत्र दक्षिणावर्तते पितृषु तर्हि भवति पितृंस्तर्हाभिगोपायति॥

शतपथ ब्राह्मणम्, २. १. ३. ३.

Sā Yatrōdaṅṅāvartatē।
Dēvēṣu Tarhi Bhavati Dēvāṃstarhābhigōpāyatyatha Yatra Dakṣiṇāvartatē Pitṛṣu Tarhi Bhavati Pitṛṃstarhābhigōpāyati॥

Śatapatha Brāhmaṇa. II. 1. 3. 3.

27. ते य एव मेतद्विदुः ये चामी अरण्ये श्रध्धाग्ंसत्यमुपासते तेर्चिरभि सम्भवन्ति अर्चिषोहरह्न अपूर्व माणपक्ष मा पूर्यमाण पक्षाद्यान् षण्मासानुदङ्ङादित्य एति मासेभ्यो देवलोकं देवलोका दादित्य मादित्याद्वैद्युतं तान्वैद्युता त्पुरुषो मानस एत्य ब्रह्मलोकागमयति तेषु ब्रह्मलोकेषु पराः परावतो वसन्ति तेषां न पुनरावृत्तिः॥

बृहदारण्यक उपनिषद्, ८. २. १५.

Tē Ya Ēva Mētadviduḥ Yē Cāmī Araṇyē Śradhdhāgṃsatyamupāsatē Tērcirabhi Sambhavanti Arciṣōharahna Apūrva Māṇapakṣa Mā Pūryamāṇa Pakṣādyān Ṣaṇmāsānudaṅṅāditya Ēti Māsēbhyō Dēvalōkaṃ Dēvalōkā Dāditya Mādityādvaidyutaṃ Tānvaidyutātpuruṣō Mānasa Ētya Brahmalōkāngamayati Tēṣu Brahmalōkēṣu Parāḥ Parāvatō Vasanti Tēṣāṃ Na Punarāvṛttiḥ।

Brhadāraṇyaka Upaniṣad, VIII. 2. 15.

अथ ये यज्ञेन दानेन तपसा लोकाञ्जयन्ति ते धूममभिसम्भवन्ति धूमाद्रात्रिग्ंरात्रेरपक्षीयमाण पक्ष मपक्षीयमाण पक्षाद्यान् षण्मासान् दक्षिणामादित्य एति मासेभ्यः पितृलोकं पितृलोकाच्चन्द्रं ते चन्द्रं प्राप्यान्नं भवन्ति तागंस्तत्र देवा यथा सोमग्ंराजानमाप्याय स्वा पक्षीयस्वे त्येवमेनागंस्तत्र भक्षयन्ति तेषां

यदा तत्पर्यवैत्यथैममेवा काशमभि निष्पद्यन्ते आकाशाद्वायुं वायोर्वृष्टिं वृष्टेः पृथवीं ते पृथिवीं प्राप्यान्नं भवन्ति ते पुनः पुरुषाग्नौ हूयन्ते त तोयोषग्नौ जायन्ते ते लोकान् प्रत्युत्थयिनस्त एव मेवानुपरिवर्तन्तेथ य एतौ पन्थानौ न विदुस्ते कीटाः पतंगा यदिदं दन्दशूकम्॥१६॥

Atha Yē Yajñēna Dānēna Tapasā Lōkāñjayanti Tē Dhūmamabhisambhavanti Dhūmādrātrigṃ Rātrērapakṣīyamāṇa Pakṣa Mapakṣīyamāṇa Pakṣādyān Ṣaṇmāsān Dakṣiṇāmāditya Ēti Māsēbhyaḥ Pitṛlōkaṃ Pitṛlōkāccandraṃ Tē Candraṃ Prāyānnaṃ Bhavanti Tāgṃ Statra Dēvā Yathā Sōmagṃ Rājānamāpyāya Svā Pakṣīyasvē Tyēvamēnāgnṃ Statra Bhakṣayanti Tēṣāṃ Yadā Tatparyavai-tyathēmamēvā Kāśamabhi Niṣpadyantē Ākāśādvāyuṃ Vāyōrvṛṣṭiṃ Vṛṣṭēḥ Pṛthivīṃ Tē Pṛthivīṃ Prāpyānnaṃ Bhavanti Tē Punaḥ Puruṣāgnau Hūyantē Ta Tōyōṣāgnau Jāyantē Tē Lōkān Pratyuttha-yinasta Ēva Mēvānuparivartantētha Ya Ētau Panthānau Na Vidustē Kīṭāḥ Pataṅgā Yadidaṃ Dandaśūkam//16//

28. इन्द्र सोमं पिब ऋतुनात्वा विशन्त्विन्दवः।
मत्सरासस्तदोकसः॥

ऋग्वेद, १. १५. १.

Indra Sōmaṃ Piba Ṛtunātvā Viśantvindavaḥ/
Matsarāsastadōkasaḥ/

Ṛgvēda, I. 15. 1.

मरुतः पिबत ऋतुना पोत्राद् यज्ञं पुनीतन।
यूयं हि ष्ठा सुदानवः॥२॥

Marutaḥ Pibata Ṛtunā Pōtrād Yajñaṃ Punītana/
Yūyaṃ Hi Ṣṭhā Sudānavaḥ//2//

अभि यज्ञं गृणीहि नो ग्नावो नेष्टः पिब ऋतुना।
त्वं हि रत्नधा असि॥३॥

Abhi Yajñaṃ Gṛṇīhi Nō Gnāvō Nēṣṭaḥ Piba Ṛtunā/
Tvaṃ Hi Ratnadhā Asi//3//

अग्ने देवाँ इहा वह सादया योनिषु त्रिषु।
परि भूष पिब ऋतुना॥४॥

Agnē Dēvām Ihā Vaha Sādayā Yōniṣu Triṣu/
Pari Bhūṣa piba Ṛtunā//4//

ब्राह्मणादिन्द्र राधसः पिबा सोममृतूँरनु।
तवेद्धि सख्यमस्तृतम्॥५॥

Brāhmaṇādindra Rādhasaḥ Pibā Sōmamṛtūmranu/
Tavēddhi Sakhyamastṛtam//5//

युवं दक्षं धृतव्रत मित्रावरूण दूलभम्।
ऋतुना यज्ञमाशाथे॥६॥
Yuvaṃ Dakṣaṃ Dhṛtavrata Mitrāvaruṇa Dūḷabham/
Ṛtunā Yajñamāśthē//6//
द्रविणोदा द्रविणसो ग्रावहस्तासो अध्वरे।
यज्ञेषु देवमीळ्ठते॥७॥
Draviṇōdā Draviṇasō Grāvahastāsō Adhvarē/
Yajñēṣu Dēvamīḷatē//7//
द्रविणोदा ददातु नो वसूनि यानि शृण्विरे।
देवेषु ता वनामहे॥८॥
Draviṇōdā Dadātu Nō Vasūni Yāni Śṛṇvirē/
Dēvēṣu Tā Vanāmahē//8//
द्रविणोदाः पिपीषति जुहोत प्र च तिष्ठत।
नेष्ट्रादृतुभिरिष्यत॥९॥
Draviṇōdāḥ Pipīṣati Juhōta Pra Ca Tiṣṭhata/
Nēṣṭrādṛtubhiriṣyata//9//
यत् त्वा तुरीयमृतुभिर्द्रविणोदो यजामहे।
अध स्मा नो ददिर्भव॥१०॥
Yat Tvā Turīyamṛtubhirdraviṇōdō Yajāmahē/
Adha Smā Nō Dadirbhava//10//
अश्विना पिबतं मधु दीद्यग्नी शुचिव्रता।
ऋतुना यज्ञवाहसा॥११॥
Aśvinā Pibataṃ Madhu Dīdyagnī Śucivratā/
Ṛtunā Yajñavāhasā//11//
गार्हपत्येन सन्त्य ऋतुना यज्ञनीरसि।
देवान् देवेयते यज॥१२॥
Gārhapatyēna Santya Ṛtunā Yajñanīrasi/
Dēvān Dēvēyatē Yajā//12//

29. शतं जीव शरदो वर्धमानः शतं हेमन्ताञ्छतमु नसन्तान्।
शतमिन्द्राग्नी सविता बृहस्पतिः शतायुषा हविषेमं पुनर्दुः॥

ऋग्वेद, १०.१६१. ४.

Śataṃ Jīva Śaradō Vardhamānaḥ Śataṃ Hēmantāñchatamu Nasantān/
Śatamindrāgnī Savitā Bṛhaspatiḥ Śatāyuṣā Haviṣēmaṃ Punarduḥ//

Ṛgvēda, X. 161. 4.

यत् पुरुषेण हविषा देवा यज्ञमतन्वत।
वसन्तो अस्यासीदाज्यं ग्रीष्म इध्मः शरद्धविः॥

ऋग्वेद, १०. ९०. ६.

Yat Puruṣēṇa Haviṣā Dēvā Yajñamatanvata l
Vasantō Asyāsīdājyaṃ Grīṣma Idhmaḥ Śaraddhaviḥ ll
Ṛgvēda, X. 90. 6.

30. Ibid.
31. यदीमेनाँ उशतो अभ्यवर्षीत् तृष्यावतः प्रावृष्यागतायाम्।
अक्खलीकृत्या पितरं न पुत्रो अन्यो अन्यमुप वदन्तमेति॥
ऋग्वेद, ७. १०३. ३.

Yadīmēnām Uśatō Abhyavarṣīt Tṛṣyāvatataḥ Prāvṛṣyāgatāyām l
Akkhalīkṛtyā Pitaraṃ Na Putrō Anyō Anyamupa Vadantamēti ll
Ṛgvēda, VII. 103. 3.

32. यः शम्बरं पर्वतेषु क्षियन्तं चत्वारिंशयां शरद्यन्वविन्दत्।
ओजायमानं यो अहिं जघान दानुं शयानं स जनास इन्द्रः॥
ऋग्वेद, २. १२. ११.

Yaḥ Śambaraṃ Parvatēṣu Kṣiyantaṃ Catvāriṃśaṃ Śaradyanvavindat l
Ōjāyamānaṃ Yō Ahiṃ Jaghāna Dānuṃ Śayānaṃ Sa Janāsa Indraḥ ll
Ṛgvēda, II. 12. 11.

वि ये दधुः शरदं मासमादहर्यज्ञमक्तुं चादृचम्।
अनाप्यं वरूणो मित्रो अर्यमा क्षत्रं राजान आशत॥
ऋग्वेद, ७. ६६. ११.

Vi Yē Dadhuḥ Śaradaṃ Māsamādaharyajñamaktuṃ Cādṛcam l
Anāpyaṃ Varuṇō Mitrō Aryamā Kṣatraṃ Rājāna Āśata ll
Ṛgveda, VII. 66. 11.

शतं जीव शरदो वर्धमानः शतं हेमन्ताञ्छतमु नसन्तान्।
शतमिन्द्राग्नी सविता बृहस्पतिः शतायुषा हविषेमं पुनर्दुः॥
ऋग्वेद, १०.१६१.४.

Śataṃ Jīva Śaradō Vardhamānaḥ Śataṃ Hēmantāñchatamu Nasantān l
Śatamindrāgnī Savitā Bṛhaspatiḥ Śatāyuṣā Haviṣēmaṃ Punarduḥ ll
Ṛgvēda, X. 161. 4.

33. शतं जीव शरदो वर्धमानः शतं हेमन्ताञ्छतमु नसन्तान्।
शतमिन्द्राग्नी सविता बृहस्पतिः शतायुषा हविषेमं पुनर्दुः॥
ऋग्वेद, १०. १६१. ४.

Śataṃ Jīva Śaradō Vardhamānaḥ Śataṃ Hēmantāñchatamu Nasantān l
Śatamindrāgnī Savitā Bṛhaspatiḥ Śatāyuṣā Haviṣēmaṃ Punarduḥ ll
Ṛgvēda, X. 161. 4.

34. ग्रीष्मो हेमन्तः शिशिरो वसन्तः शरद् वर्षाः स्विते नो दधात।
आ नो गोषु भजता प्रजायां निवात इद् वः शरणे स्याम॥

अथर्ववेद, ६. ५५. २.

Grīṣmō Hēmantaḥ Śiśirō Vasantaḥ Śarad Varṣāḥ Svitē Nō Dadhāta I
Ā No Gōṣu Bhajatā Prajāyāṃ Nivāta Id Vaḥ Śaraṇē Syāma II

Atharvavēda, VI. 55. 2.

35. पञ्चार्तवः संवत्सरस्य हेमंतंशिशिरयोः समानेन।

ऐतरेय ब्राह्मणम्, १. १.

Pancārtavaḥ Saṃvatsarasya Hēmantaṃśiśirayōḥ Samānēna I

Aitarēya Brāhmaṇa I. 1.

36. Maitrāyaṇīya Saṃhitā, I. 7. 3.

37. वसन्तो ग्रीष्मो वर्षाः। ते देवा ऋतवः। शरद् हेमन्तः शिशिरस्ते पितरो य एवा पूर्यतेर्द्ध मासः स देवा योपक्षीयते स पितरोहरेव देवा रात्रिः पितरः पुनरह्नः पूर्वाह्णो देवा अपराह्नः पितरः।

ते वा एत ऋतवः। देवाः पितरः स यो हैवं व्विद्वान्देवाः पितर इति ह्वयत्या हास्य देवा देवहूयं गच्छन्त्या पितरः पितृहूयमवन्ति हैनं देवा देवहूयेवन्ति पितरः पितृहूये य एवं व्विद्वान्देवाः पितर इति ह्वयति॥

सा यत्रोदङ्ङावर्तते। देवेषु तर्हि भवति देवांस्तर्हाभिगोपायत्यथ यत्र दक्षिणावर्तते पितृषु तर्हि भवति पितृंस्तर्हाभिगोपायति॥

सा यत्रोदङ्ङावर्तते तह्र्यग्नी आदधीतापहतपाप्मानो देवा अप पाप्मानँ हतेमृता देवा नामृतत्वस्याशास्ति सर्व्वमायुरेति यस्तह्र्याधत्तेथ यत्र दक्षिणावर्तते यस्तह्र्यधत्तेनपहत पाप्मानः पितरो न पाप्मानमपहते मर्त्याः पितरः पुरा हायुषो म्रियते यस्तह्र्याधत्ते॥

ब्रह्मैव व्वसन्तः क्षत्रं ग्रीष्मो व्विड़ेव व्वर्षास्तस्माद्ब्रह्मणो व्वसन्त आदधीत ब्रह्म हि व्वसन्तस्मत्क्षत्रियो ग्रीष्मादधीत क्षत्रम् हि ग्रीष्म तस्माद्वैश्यो व्वर्षास्वादधीत विड्ढि व्वर्षाः॥

स युः कामयेत। ब्रह्मव्वर्चसी स्यामिति व्वसन्ते स आदधीत ब्रह्म वै व्वसन्तो ब्रह्मवर्चसी हैव भवति॥

शतपथ ब्राह्मणम्, २. १. ३. १–६.

Vasantō Grīṣmō Varṣāḥ. Tē Dēvā Ṛtavaḥ I Śarad Hēmantaḥ Śiśirastē Pitarō Ya Ēvā Pūryatērddha Māsaḥ Sa Dēvā Yōpakṣīyatē Sa Pitarōharēva Dēvā Rātriḥ Pitaraḥ Punarahnaḥ Pūrvāhnō Dēvā Aparāhnaḥ Pitaraḥ I

Tē Vā Ēta Ṛtavaḥ I Dēvāḥ Pitaraḥ Sa Yō Haivaṃ Vvidvāndēvāḥ Pitara Iti Hvayatyā Hāsya Dēvā Dēvahūyaṃ Gacchantyā Pitaraḥ Pitṛhūyamavanti Hainaṃ Dēvā Dēvahūyēvanti Pitaraḥ Pitṛhūyē Ya Evaṃ Vvidvāndēvāḥ Pitara Iti Hvayati I

Sā Yatrōdaṅṅāvartatē। Dēvēṣu Tarhi Bhavati Dēvāṃstarhābhigōpāyatyatha Yatra Dakṣiṇāvartatē Pitṛṣu Tarhi Bhavati Pitṛṃstarhābhigōpāyati॥
Sā Yatrōdaṅṅāvartatē Tahryagnī Ādadhītāpahatapāpmānō Dēvā Apa Pāpmānam Hatēmṛtā Dēvā Nāmṛtatvasyāśāsti Sarvvamāyurēti Yastahryādhattētha Yatra Dakśiṇāvartatē Yastahryādhattēnapahata Pāpmānaḥ Pitaro Na Pāpmānamapahatē Martyāḥ pitaraḥ Purā Hāyuṣō Mriyatē Yastahryādhattē॥
Brahmaiva Vvsantaḥ Kṣatraṃ Grīṣmō Vviḍēva Vvrṣāstasmādbrahmaṇō Vvasanta Ādadhīta Brahma Hi Vvasantasmātksatriyō Grīṣmādadhīta Kṣatram Hi Grīṣma Tasmādvaiśyō Vvarṣāsvādadhīta Viḍḍhi Vvarṣāḥ॥
Sa Yuḥ Kāmayēta. Brahmavvarcasī Syāmiti Vvasantē Sa Ādadhīta Vai Vvasantō Brahmavarcasī Haiva Bhavati॥

Śatapatha Brāhmaṇa. II. 1. 3. 1-6.

38. अहं विवेच पृथिवीमुत द्यामहमृतूंरजनयं सप्थ साकम्।
अहं सत्यमनृतं यद् वदाम्यहं दैवीं परि वाचं विशश्च॥

अर्थवेद, ६. ६१. २.

Ahaṃ Vivēca Pṛthivīmuta Dyāmahamṛtūṃrajanayaṃ Saptha Sākam।
Ahaṃ Satyamanṛtaṃ Yad Vadāmyahaṃ Daivīṃ Pari Vācaṃ Viśaśca॥

Atharvavēda, Vl. 61. 2.

39. पर्यू षु प्र धन्वा वाजसातये परि वृत्राणि शक्षणिः।
द्विषस्तदध्यर्णवेनेयसे सनिस्रसो नामासि त्रयोदशो मास इन्द्रस्य गृहः॥

अथर्ववेद, ५. ६. ४.

Paryū Ṣu Pra Dhanvā Vājasātayē Pari Vṛtrāṇi Śakṣaṇiḥ।
Dviṣastadadhyarṇavēneyasē Sanisrasōnāmāsi Trayōdaśō Māsa Indrasya Gṛhaḥ॥

Atharvavēda, V. 6. 4.

40. मधुश्च माधवश्च वासन्तिकावृतू शुक्रश्च शुचिश्च ग्रैष्मावृतू नभश्च नभस्यश्च वार्षिकावृतू इषश्चोर्जश्च शारदावृतू सहश्च सहस्यश्च हैमन्तिकावृतू तपश्च तपस्यश्च शैशिरावृतू अग्नेरन्तः श्लेषोसि कल्पेतां द्यावापृथिवी कल्पन्तामाप ओषधीः कल्पन्तामग्नयः पृथङ्ममज्येष्ठ्याय सव्रताः।

तैत्तिरीय संहिता, ४. ४. ११. १.

Madhuśca Mādhavaśca Vāsantikāvṛtū Śukraśca Śuciśca Graiṣmāvṛtū Nabhaśca Nabhasyaśca Vārṣikāvṛtū Iṣaścōrjaśca Śāradāvṛtū Sahaśca Sahasyaśca Haimantikāvṛtū Tapaśca Tapasyaśca Śaiśirāvṛtū Agnērantaḥ Ślēṣōsi Kalpētāṃ Dyāvāpṛthivī Kalpantā-

māpa Ōṣadhīḥ Kalpantāmagnayaḥ Pṛthaṅmama Jyēṣṭhyāya Savratāḥ. Taittirīya Saṃhitā, IV. 4. 11. 1.

41. मुखं वा एतद्ऋतूनाम् यद्वसन्तः।

तैत्तिरीय ब्राह्मणम्, १. १. २. ६.

Mukhaṃ Vā Ētadṛtūnām Yadvasantaḥ.

Taittirīya Brāhmaṇa, 1. 1. 2. 6.

42. op. cit. Sat. Br., 2. 1. 3. 1-5.
43. Prof. Renou's article on 'Vēdic Ṛtus' in 'Indian Culture', Vol. XV. pp. 21-26.
44. C.I.I., Vol. I. p. 175.
45. E.I., Vol. XX. p. 16. 'Sirivīra Puriṣadatasa Sava 6 Vā Pa 6 Diva 10'; at p. 21. Saṃvaccharaṃ Aṭṭāra Saṃ 108 Hēmanta Pakhaṃ Chataṃ 6 Dava Pañcamaṃ.
46. E.I., Vol. I. 15. p. 252. 'Hēmanta Pakṣē Tṛtīyē Trayōdaśyām'.
47. E.I., Vol. XVII. p. 11 in sa (Saṃvat) 8, gri 4, di 5.
48. Vide Luder's list no. 1202. samvachhare 20, Vāsa Pakhaṃ 8, Divasaṃ 10.
49. ब्रह्मं दैवं मानुषं च पित्र्यं सौरं च सावनम्।
चान्द्रमार्क्षं गुरोर्मानमिति मानानि वै नव।
एषां तु नवमानाम् व्यवहारोत्र पञ्चभिः।
तेषां पृथक् पृथक् कार्यं वक्ष्यते व्यवहारतः॥

नारद संहिता, ३. १-२.

Brahmaṃ Daivaṃ Mānuṣaṃ Ca Pitryaṃ Sauraṃ Ca Sāvanam।
Cāndramārkṣaṃ Gurōrmānamiti Mānāni Vai Nava।
Ēṣāṃ Tu Navamānām Vyavahārōtra Pañcabhiḥ।
Tēṣāṃ Pṛthak Pṛthak Kāryaṃ Vakṣyatē Vyavahārataḥ॥

Nārada Saṃhitā, III. 1-2.

50. Hēmādri on Kāla, p. 9.
51. Mādhava, Kālanirṇaya-Kārika, 11-12.
52. नाडीषष्ट्या तु नाक्षत्र महोरात्रं प्रकीर्तितम्।
तत्रिंशता भवे न्मासः सावनोर्कोदयै स्तथा॥
ऐन्दव स्थिथिभि स्तद्वत् संक्रान्त्या सौर उच्यते।
मासैर्द्वादशभिर्वर्षं दिव्यं तदः रूच्यते॥

सूर्य सिद्धांतम्, १. १३-१४.

Nāḍīṣaṣṭyā Tu Nākṣatra Mahōrātraṃ Prakīrttam।
Tatriṃśatā Bhavē Nmāsaḥ Sāvanōrkōdayai Stathā॥
Aindava Sthithibhi Stadvat Saṅkrāntyā Saura Ucyatē।
Māsairdvādaśabhirvarṣaṃ Divyaṃ Tadaḥ Rucyatē॥

Sūrya Siddhānta, I. 13-14.

53. Vide 'Indian Culture', Vol. VIII. pp. 114-116.
54. Vāyu Purāṇa, Chapter 50. verse 202.
55. Gupta Inscriptions, p. 79 ff.
56. E.I., Vol. XIX. p. 1. at p. 15; C.I.I., Vol. II. p. 145. 'Sam 10 1 Āṣāḍha Māsasa 20, Uttaraphalguṇa'.
57. Indian Historical Quarterly (I.H.Q.), Vol. XI. p. 300 at 302.
58. E.I., Vol. XXVII. p. 319 at p. 325. 'Tasambatsarē Sapta Pañcāśattamē Kārtika Pūrṇamāsyām Ankēna Saṃvat 57, Kārtīka-Dina 30'.
59. E.I., Vol. XXIV. pp. 1-8, the Bajaur Casket Inscription of Minadra.
60. E.I., Vol. XIX. p. 65 (seven incriptions from Mathura. at p. 66 Saṃ 22 Gṛ 2 Di 30).
61. E.I., Vol. VIII. at p. 176. 'Saṃ 3 Hē 3 Di 22' Sāraṇāth Inscription of Kaniṣka.
62. Gupta Inscriptions, p. 121 and 125.
63. क्षयं संवत्सराणाम् च मासानाम् च क्षयं तथा।
पक्ष क्षयं तथा दृष्ट्वा दिवसानाम् च संक्षयम्।
क्षयं वृद्धिं च चन्द्रस्य दृष्ट्ववा। प्रत्यक्ष तस्तथा।।

महाभारतम्, शांतिपर्वम्, ३०१. ४६-४७.

Kṣayaṃ Saṃvatsarāṇām Ca Māsānām Ca Kṣayaṃ Tathā।
Pakṣa Kṣayaṃ Tathā Dṛṣṭvā Divasānām Ca Saṅkṣayam।
Kṣayaṃ Vṛddhiṃ Ca Candrasya Dṛṣṭvā। Pratyakṣa Tastathā।।

Mahābhārata, Śāntiparvan, 301. 46-47.

64. तं त्रयोदशन्मासादक्रीणन् तस्मात् त्रयोदशो मासो नानुविद्यते न वै सोमविक्रय्यनुविद्यते पापो हि सोमविक्रयी।

ऐतरेय ब्राह्मणम्, ३. १.

Taṃ Trayōdaśanmāsādakrīṇan Tasmāt Trayōdaśō Māsō Nānuvidyatē Na Vai Sōmavikrayyanuvidyatē Pāpō Hi Sōmavikrayī/

Aitarēya Brāhmaṇa, III. 1.

65. मधुश्च माधवश्च शुक्रश्च शुचिश्च नभश्च
नभस्यश्चेषश्चोर्जश्च सहश्च सहस्यश्च तपश्च
तपस्यश्चोपयामगृहीतोसि सँसर्पोस्यँहस्पत्याय त्वा।।

तैत्तिरीय संहिता, १. ४. १४.

Madhuśca Mādhavaśca Śukraśca Śuciśca Nabhaśca
Nabhasyaścēṣaścōrjaśca Sahaśca Sahasyaśca Tapaśca
Tapasyaścōpayāmagṛh ītōsi Somsarpōsyamhaspatyāya Tvā।।

Taittirīya Saṃhitā, 1. 4. 14.

66. Hēmādri on Kāla, p. 28.
67. पलालानुपलालौ शर्कुम् कोकम् मलिम्लुचम् पलीजकम्।
आश्रेषम् वव्रिवाससमृक्षग्रीवम् प्रमीलिनम्।।

अथर्ववेद, ८. ६. २.

Palālānupalālau Śarkum Kōkam Malimlucam Palījakam।
Āśrēṣam Vavrivāsasamṛkṣagrīvam Pramīlinam॥
Atharvavēda, VIII. 6. 2.

68. Padma Purāṇa, VI. 64.
69. अग्न्यादधेयम् प्रतिष्ठाम् च यज्ञ दान व्रतानि च।
वेदव्रत वृषोत्सर्ग चूढाकरणमेखलाः।
मांगल्यमभिषेकं च मलमासे विवर्जयेत्॥
अग्नि पुराणम्, १७५. २९-३०.
Agnyādadhēyam Pratiṣṭhām Ca Yajña Dāna Vratāni Ca।
Vēdavrata Vṛṣōtsarga Cūḍhākaraṇamēkhalāḥ।
Māṅgalyamabhiṣēkaṃ Ca Malamāsē Vivarjayēt॥
Agni Purāṇa, 175. 29-30.
70. I.A., I, Vol. XIV. P. 351.
71. E.I., Vol. XVI. p. 236 at 238.
72. Luder's list No. 965 and E.I., Vol. VIII. p. 36 at p. 40.
73. Gupta Inscritions, at p. 258.
74. E.I., Vol. VIII. at p. 82.
75. C.I.I., Vol. IV. p. 22, 24 (Vaiśākha-Śuddha-Trayōdaśyām).
76. E.I., Vol. XVII. p. 110.
77. तेषाम् पूर्वपक्षे सुत्या संपद्यते।
ताण्ड्य महा ब्राह्मण, ५. ९. १४.
Tēṣām Pūrvapakṣē Sutyā Sampadyatē।
Tāṇḍya Mahā Brāhmaṇa, V, 9. 14.
तं पूर्वपक्षे याजयेत् वसीयानेव भवति।
यं कामयेत् पापीयान्स्यादिति तमपरपक्षे याजयेत्।
पापीयनेव भवति। तस्मात्पूर्वपक्षोपरपक्षात् कारूण्यतरः।
तैत्तिरीय ब्राह्मणम्, २. २. ३. १.
Taṃ Pūrvapakṣē Yājayēt Vasīyānēva Bhavati।
Yaṃ Kāmayēt Pāp īyānsyāditi Tamaparapakṣē Yājayēt।
Pāpīyanēva Bhavati। Tasmātpūrvapakṣōparapakṣāt Kāruṇyataraḥ।
Taittirīya Brāhmaṇa, II, 2. 3. 1.
78. अथ य दु चैवास्मि ञ्छव्यं कुर्वन्ति यदिच नार्चिष मेवाभिसम्भवन्त्यर्चिषोहरह्न आपूर्यमाण पक्षमापूर्यमाण पक्षाद्यान् षडुदञ्जेति मासांस्तान्मासेभ्य स्संवत्सरं संवत्सरादादित्य मादित्या च्चन्द्रमसं चन्द्रमसोविद्युतं तत्पुरुषोमानवः। सएनां ब्रह्म गमय त्येष देवपथो ब्रह्मपथ एतेन प्रतिपद्यमाना इमं मानव मावर्तं नावर्तन्ते नावर्तन्ते।
छांदोग्य उपनिषद्, ४. १५. ५.

Atha Ya Du Caivāsmi Ñchavyaṃ Kurvanti Yadica Nārciṣa Mēvā Abhisambhavantyaciṣōharahna Āpūryamāṇa Pakṣamāpūryamāṇa Pakṣādyān Ṣaḍudajjēti Māsāṃstānmāsēbhya Ssaṃvatsaraṃ Saṃvatsarādāditya Mādityā Ccandramasaṃ Candramasōvidyutaṃ Tatpuruṣōmānavaḥ। Saēnāṃ Brahma Gamaya Tyēṣa Dēvapathō Brahmapatha Ētēna Pratipadyamānā Imaṃ Mānava Māvartaṃ Nāvartantē Nāvartantē.

Chāndōgya Upaniṣad, IV. 15. 5.

त द्य इत्थं विदः ये चेमेरण्ये श्रद्धा तप इत्युपासते तेर्चिष मभिसंभवंत्यर्चिषोहरह्न आपूर्यमाणपक्ष मापूर्यमाणपक्षाद्यान् षड़ुदङ्ङेति मासां स्तान्।

छांदोग्य उपनिषद्, ५. १०. १.

Ta Dya Itthaṃ Viduḥ Yē Cēmēraṇyē Śraddhā Tapa Ityupāsatē Tērciṣa Mabhisambhavantyarciṣōharahna Āpūryamāṇa Pakṣa Māpūryamāṇapakśādyān Ṣaḍudajñjñēti Māsāṃ Stān।

Chāndōgya Upaniṣad, V. 10. 1.

79. ते य एव मेतद्विदुः ये चामी अरण्ये श्रधधागं् सत्यमुपासते तेर्चिरभि सम्भवन्ति अर्चिषोहरह्न आपूर्यमाणपक्ष मापूर्यमाणपक्षाद्यान् षण्मासानुदङ्ङादित्य एति मासेभ्यो देवलोकं देवलोका दादित्य मादत्याद्वैद्युतं तान्वैद्युता त्पुरुषो मानस एत्य ब्रह्मलोकान्गमयति तेषु ब्रह्मलोकेषु पराः परावतो वसन्ति तेषां न पुनरावृत्तिः।

बृहदारण्यक उपनिषद्, ८. २. १५.

Tē Ya Ēva Mētadviduḥ Yē Cāmī Araṇyē Śradhdhāgṃ Satyamupāsatē Tērcirabhi Sambhavanti Arciṣōharahna Āpūryamāṇapakṣa Māpūryamāṇapakṣādyān Ṣaṇmāsānudaṅṅāditya Ēti Māsēbhyō Dēvalōkaṃ Dēvalōkā Dāditya Mādityādvaidyutaṃ Tānvaidyutā Tpuruṣō Mānasa Ētya Brahmalōkāngamayati Tēṣu Brahmalōkēṣu Parāḥ Parāvatō Vasanti Tēṣāṃ Na Punarāvṛttiḥ।

Bṛhadāraṇyaka Upaniṣad, VIII, 2. 15.

अथ ये यज्ञेन दानेन तपसा लोकाञ्जयन्ति ते धूम मभिसम्भवन्ति धूमाद्रात्रिगं् रात्रेरपक्षीयमाण पक्ष मपक्षीयमाण पक्षाद्यान् षण्मासान् दक्षिणामादित्य एति मासेभ्यः पितृलोकं पितृलोकाच्चन्द्रं ते चन्द्र प्राप्यान्नं भवन्ति तागं् स्तत्र देवा यथा सोमगं् राजानमाप्याय स्वा पक्षीयस्वे त्येवमेनागं् स्तत्र भक्षयन्ति तेषां यदा तत्पर्यवैत्यथेममेवा काशमभि निष्पद्यन्ते आकाशाद्वायुं वायोर्वृष्टिं वृष्टेः पृथिवीं ते पृथिवीं प्राप्यान्नं भवन्ति ते पुनः पुरुषाग्नौ हूयन्ते त तोयोषाग्नौ जायन्ते ते लोकान् प्रत्युत्थायिनस्त एव मेवानुपरिवर्तन्तेथ य एतौ पन्थानौ न विदुस्ते कीटाः पतंगा यदिदं दन्दशूकम्॥१६॥

Atha Yē Yajñēna Dānēna Tapasā Lōkāñjayanti Tē Dhūma Mabhisambhavanti Dhūmādrātrigṃ Rātrērapakṣīyamāṇa Pakṣa

Mapakṣīyamāṇa Pakṣādyān Ṣaṇmāsān Dakṣiṇāmāditya Ēti Māsēbhyaḥ Pitṛlōkaṃ Pitṛōkāccandraṃ Tē Candraṃ Prāpyānnaṃ Bhavanti Tāgṃ Statra Dēvā Yathā Sōmagṃ Rājānamāpyāya Svāpakṣīyasvē Tyēvamēnāgṃ Statra Bhakṣayanti Tēṣāṃ Yadā Tatparyavaityathēmamēvā Kāśamabhi Niṣpadyantē Ākāśādvāyuṃ Vāyōrvṛṣṭhiṃ Vṛṣṭēḥ Pṛthivīṃ Tē Pṛthivīṃ Prāpyānnaṃ Bhavanti Tē Punaḥ Puruṣāgnau Hūyantē Ta Tōyōṣāgnau Jāyantē Te Lōkān Pratyutthāyinasta Ēva Mēvānuparivartantētha Ya Ētau Panthānau Na Vidustē Kīṭāḥ Pataṅgā Yadidaṃ Dandaśūkam//16//

80. Ibid., III.1. 5.
81. Pāṇini, V. 2. 25.
82. Asiatic Researches, Vol. III, Article on 'Lunar year of the Hindus', by the President.
83. **सिनीवालि पृथुष्टुके या देवानामसि स्वसा।**
जुषस्व हव्यमाहुतं प्रजां देवि दिदिड्ढि नः॥
या सुबाहुः स्वङ्गुरिः सुषूमा बहुसूवरी।
तस्यै विश्पत्न्यै हविः सिनीवाल्यै जुहोतन॥

ऋग्वेद, २. ३२. ६-७.

Sinīvāli Pṛthuṣṭukē Yā Dēvānāmasi Svasā/
Juṣasva Havyamāhutaṃ Prajāṃ Dēvi Didiṅḍhi Naḥ//
Yā Subāhuḥ Svaṅguriḥ Suṣūmā Bahusūvarī/
Tasyai Viśpatnyai Haviḥ Sinīvāiyai Juhōtana/

Ṛgvēda, II. 32. 6-7.

84. **राकामहं सुहवां सुष्टुती हुवे शृणोतु नः सुभगा बोधतु त्मना।**
सीव्यत्वपः सीच्याच्छिद्यमानया ददातु वीरं शतदायमुक्थ्यम्॥

ऋग्वेद, २. ३२. ४.

Rākāmahaṃ Suhavāṃ Suṣṭutī Huvē Śṛṇōtu Naḥ Subhagā Bōdhatu Tmanā/
Sīvyatvapaḥ Sīcyācchidyamānayā Dadātu Vīraṃ Śatadāyamukthyam// Ṛgvēda, II. 32. 4.

85. **असुनीते पुनरस्मासु चक्षुः पुनः प्राणमिह नो धेहि भोगम्।**
ज्योक् पश्येम सूर्यमुच्चरन्त मनुमते मृळया नः स्वस्ति॥

ऋग्वेद, १०. ५९. ६.

Asunītē Punarasmāsu Cakṣuḥ Punaḥ Prāṇamiha Nō Dhēhi Bhōgam/
Jyōk Paśyēma Sūryamuccaranta Manumatē Mṛlayā Naḥ Svasti//

Ṛgvēda, X. 59. 6.

सोमस्य राज्ञो वरुणस्य धर्मणि बृहस्पतेरनुमत्या उ शर्मणि।
तवाहमद्य मघवन्नुपस्तुतौ धातर्विधातः कलशाँ अभक्षयम्॥

ऋग्वेद, १०. १६७. ३.

Sōmasya Rājñō Varuṇasya Dharmaṇi Bṛhaspatēranumatyā U Śarmaṇi ।
Tavāhamadya Maghavannupastutau Dhātarvidhātaḥ Kalaśām Abhakṣayam ।।

Ṛgvēda, X. 167. 3.

86. प्रजापतिरनुमतिः सिनीवाल्याचीक्लृपत्।
स्त्रैषूयमन्यत्र दधत् पुमांसमु दधदिह॥

अथर्ववेद, ६, ११, ३.

Prajāpatiranumatiḥ Sinīvālyācīkiṛpat ।
Straiśūyamanyatra Dadhat Pumāṃsamu Dadhadiha ।।

Atharvavēda, VI. 11. 3.

87. तथा तिथि नक्षत्र पर्वसमवाये।

गोभिल गृह्य सूत्रम्, १. १. १३.

Tathā Tithi Nakṣatra Parvasamavāyē ।

Gōbhila Gṛhya Sūtra, I. 1. 13.

88. Kauṣītakī Gṛhya Sūtra, I. 25; V. 2.
89. कमलजविधातृहरियमशशाङ्कषड्वक्त्रवसुभुजगाः।
धर्मेशसवितृमन्मथकलयो विश्वे च तिथिपतयः॥
पितरोमावास्यायां सञ्ज्ञासदृशाश्च तैः क्रियाः कार्याः।
नन्दा भद्रा विजया रिक्ता पूर्णा च तास्त्रिविधाः॥

बृहत्संहिता, ९९. १-२.

Kamalajavidhātṛhariyamaśaśāṅkaṣaṅvaktravasubhujagāḥ ।
Dharmēśasavitṛmanmathakalayō Viśvē Ca Tithipatayaḥ ।।
Pitarōmāvāsyāyāṃ Sañjñāsadṛśāśca Taiḥ Kriyāḥ Kāryāḥ ।
Nandā Bhadrā Vijayā Riktā Pūrṇā Ca Tāstrividhāḥ ।।

Bṛhatsaṃhitā, 99. 1-2.

90. Skanda Purāṇa, I. 133. 78-82.
91. Garuḍa Purāṇa, I. 137. 16-19.
92. Nārada Purāṇa, I. 56. 133-135.
93. पितरोमावास्यायां सञ्ज्ञासदृशाश्च तैः क्रियाः कार्याः।
नन्दा भद्रा विजया रिक्ता पूर्णा च तास्त्रिविधाः॥
यत्कार्यं नक्षत्रे तद्दैवत्यासु तिथिषु तत्कार्यम्।
करणमुहूर्तेष्वपि तत् सिद्धिकरं देवतासदृशम्॥

बृहत्संहिता, ९९. २-३.

Pitarōmāvāsyāyāṃ Sañjñāsadṛśāśca Taiḥ Kriyāḥ Kāryāḥ ।
Nandā Bhadrā Vijayā Riktā Pūrṇā Ca Tāstrividhāḥ ।।

Yatkāryaṃ Nakṣatrē Taddaivatyāsu Tithiṣu Tatkāryam।
Karaṇamuhūrtēṣvapi Tat Siddhikaraṃ Dēvatāsadṛśam।।

Bṛhatsaṃhitā, 99. 2-3.

94. तिथि नक्षत्र वारादि साधनम् पुण्यपापयोः।
प्रधान गुण भावेन स्वातन्त्रेण न ते क्षमाः।।

गर्ग, quoted by तिथितत्त्व, p.४.

Tithi Nakṣatra Vārādi Sādhanam Puṇyapāpayōḥ।
Pradhāna Guṇa Bhāvēna Svātantrēṇa na tē kṣamāḥ।।

Garga, quoted by Tithi tatva, p. 4.

95. सूर्यास्त मान पर्यन्तं यस्मिन् वारे तु या तिथिः।
विन्धते सा त्वखन्डा स्यान्नयूना चेत् खन्ड संज्ञिता।।

नारद पुराणम्, १. ५६. १५४.

Sūryāsta Māna Paryantaṃ Yasmin Vārē Tu Yā Tithiḥ।
Vindhatē Sā Tvakhanḍā Syānnayūnā Cēt Khanḍa Sañjñitā।।

Nārada Purāṇa, I. 56. 154.

96. पूर्वह्णो वै देवानां मध्यन्दिनो मनुष्याणामपरह्णः पितृणाम्।

शतपथ ब्राह्मणम्, २. ४. २. ८.

Pūrvahnō Vai Dēvānāṃ Madhyandinō Manuṣyāṇāmaparahnaḥ Pitṛṇām।

Śataspatha Brāhmaṇa, II. 4. 2. 8.

97. Manu, IV. 152.

98. श्रौतं स्मार्तं व्रतं दानं यच्चान्यत् कर्म वैदिकं।
अनिर्णीतासु तिथिषु न किञ्चित्फलति द्विज।।

नारद पुराणम्, पूर्वार्ध, २९. २.

Śrautaṃ Smārtaṃ Vrataṃ Dānaṃ Yaccānyat Karma Vaidikaṃ।
Anirṇītāsu Tithiṣu Na Kincitphalati Dvija।।

Nārada Purāṇa, Pūrvārtha, 29. 2.

99. महामाघी प्रयागे तु नैमिषे फाल्गुणी तथ।
शालग्रामे महाचैत्री एताः पुण्यतमाः स्मृताः।
महद्वारे च वैशाखी ज्येष्ठी तु पुरुषोत्तमे।
आषाढी वै कनखले केदारे श्रावणी तथा।
महाभाद्री बदर्यां तु कुब्जायं च महाश्विनी।
पुष्करे कार्तिकी कन्याकुब्जे मार्गी सदा स्मृता।
अयोध्यायां महापौषी एताः स्युः सुमहाफलाः।
स्नानं ह्येतेषु कर्तव्यं मानवैः पुष्तिकांक्षभिः।

These Verses were quoted by हेमाद्रि on काल, p. ६४२.

Mahāmāghī Prayāgē Tu Naimiṣē Phālguṇī Tatha/
Śālagrāmē Mahācaitrī Ētāḥ Puṇyatamāḥ Smṛtāḥ/
Mahadvārē Ca Vaiśākhī Jyēṣṭhī Tu Puruśōttamē/
Āṣāḍhī Vai Kanakhalē Kēdārē Śrāvaṇī Tathā/
Mahābhādrī Badaryāṃ Tu Kubjāyaṃ Ca Mahāśvinī/
Puṣkarē Kārtikī Kanyākubjē Mārgī Sadā Smṛtā/
Ayōdhyāyāṃ Mahāpauṣī Ētāḥ Syuḥ Sumahāphalāḥ/
Snānaṃ Hyētēṣu Kartavyaṃ Mānavaiḥ Puṣikankṣabhiḥ/

These verses were quoted by Hēmādri on Kāla, p. 642.

100. Viṣṇu Purāṇa, III, 11. 118-119.
101. Gupta Inscriptions.
102. क्षितितनयदिवसवारो न शुभकृदिती यदि पितामहप्रोक्ते।
कुजदिनमनिष्टमिति वा कोत्र विशेषो नृदिव्यकृतेः।

बृहत्संहिता, १. ४.

Kṣititanayadivasavārō Na Śubhakṛditī Yadi Pitāmahaprōktē/
Kujadinamaniṣṭamiti Vā Kōtra Viśēṣō Nṛdivyakṛtēḥ/

Bṛhatsaṃhitā, I. 4.

103. Two of which are quoted below by way of sample:
नृपाभिषेकं नृपतिप्रयाणं कार्यं नृपदर्शनं च।
यच्चाग्निकार्यं भुवि किञ्च दुक्तं तत्सर्वमादित्यदिने प्रशस्तम्॥
स्थाप्यं च कर्म क्रतुभिश्च यूपो गृहप्रवेशो गजबन्धनं च।
ग्रमेधिवासो नगरे पुरे वा शनैश्चरे सर्वमिदं प्रशस्तम्॥

अथर्वण वेदांग ज्योतिषम्, वार प्रकरणम्, २ & ८.

Nṛpābhiṣēkaṃ Nṛpatiprayāṇaṃ Kāryaṃ Nṛpadarśanaṃ Ca/
Yaccāgnikāryaṃ Bhuvi Kiñca Duktaṃ Tatsarvamādityadinē Praśastam//
Sthāpyaṃ Ca Karma Kratubhiśca Yūpō Gṛhapravēśō Gajabandhanaṃ Ca/
Gramēdhivāsō Nagarē Purē Vā Śanaiścarē Sarvamidaṃ Praśastam//

Atharvaṇa Vēdāṅga Jyōtiṣam, Vāra Prakaranam, 2 & 8.

104. Nārada Purāṇa, I. 51. 80.
105. Matsya Purāṇa, Chapter 93. 7. ff.
106. Ibid., 253. 7.
107. अप त्ये तायवो यथा नक्षत्रा यन्त्यक्तुभिः। सूराय विश्वचक्षसे॥

ऋग्वेद, १. ५०. २.

Apa Tyē Tāyavō Yathā Nakṣatrā Yantyaktubhiḥ/ Sūrāya Viśvacakṣasē//

Ṛgveda, I. 50. 2.

108. अमी य ऋक्षा निहितास उच्चा नक्तं ददृश्रे कुह चिद् दिवेयुः।

ऋग्वेद, १. २४. १०.

Amī Ya Ṛkṣā Nihitāsa Uccā Naktaṃ Dadṛśrē Kuha Cid Divēyuḥ।

Ṛgvēda, I. 24. 10.

109. Atharvaṇa Nakṣatra Kalpa, Chapter 4 Verses 1-8.

110. अश्वियमदहनकमलजशशिशूलभृदितिजीवफणिपितरः।
योन्यर्यमदिनकृत्तवष्टृपवनशक्राग्निमित्राश्च॥
शुक्रो निरृतिस्तोयं विश्वे ब्रह्मा हरिर्वसुर्वरूणः।
अजपादोहिर्बुध्न्यः पूषा चेतीश्वरा भानाम्॥

बृहत्संहिता, ९८. ४-५.

Aśviyamadahanakamalajaśaśiśūlabhṛdaditijīvaphaṇipitaraḥ।
Yōnyaryamadinakṛttvaṣṭṛpavanaśakrāgnimitrāśca॥
Śukrō Nirṛtīstōyaṃ Viśvē Brahmā Harirvasurvaruṇaḥ।
Ajapādōhirbudhnyaḥ Pūṣā Cētīśvarā Bhānām॥

Bṛhatsaṃhitā, 98. 4-5.

111. Śatapatha Brāhmaṇa, II, 1. 2.

112. स यः कामयेत महत्प्राप्नु या मित्युदगयन आपूर्यमाण पक्षस्य पुण्या हेद्वादशाह मुपसद्व्रती भूत्वा औदुम्बरे कगूंसे चमसे वा सर्वौषधं फलानीति संभृत्य परिसमूह्य परिलिप्याग्नि मुपसमाधाय परिस्तीर्यावृताज्यगं् सगं्स्कृत्य पुगं्सा नक्षत्रेण मंथगं् संनीय जुहोति।

बृहदारण्यक उपनिषद्, ८. ३. १; छान्दोग्य उपनिषद्, ५. २. ४-९.

Sa Yaḥ Kāmayēta Mahatprāpnu Yā Mityudagayana Āpūryamāṇa Pakṣasya Puṇyā Hēdvādaśāha Mupasadvratī Bhūtvā Oudumbarē Kagṃsē Camasē Vā Sarvauṣadhaṃ Phalānīti Sambhṛtya Parisamūhya Parilipyāgni Mupasamādhāya Paristīryāvṛtājyagṃ Sagṃskṛtya Pugṃsā Nakṣatrēṇa Manthagṃ Sannīya Juhōti।

Bṛhadāraṇyaka Upaniṣad, VIII, 3. 1;
Chāndōgya Upaniṣad, V, 2. 4-9;

अधात एक धनावरोधनं यदेकधनमभिध्यायात्।
पौर्णमास्यां वामवास्यां वा शुद्ध पक्षे वा पुण्ये नक्षत्रेग्नि मुपसमाधाय परिसमूह्य।

कौषीतकी उपनिषद्, २. ३.

Adhāta Ēka Dhanāvarōdhanaṃ Yadēkadhanamabhidhyāyāt।
Paurṇamāsyāṃ Vāmāvāsyāṃ Vā Śuddha Pakṣē Vā Puṇyē Nakṣatrēgni Mupasamādhāya Parisamūhya।

Kauṣītakī Upaniṣad, II. 3.

113. उग्राण्यार्द्रा च चित्रा च विशाखा श्रवणे श्वयुक्।
क्रूराणि तु मघा स्वाती मूलं ज्येष्ठा यमस्य यत्॥

वेदांग ज्योतिष (Yajus. verse. ४२).

Ugrāṇyārdrā Ça Citrā Ca Viśākhā Śravaṇē Śvayuk/
Krūrāṇi Tu Maghā Svātī Jyēṣṭhā Mūlaṃ Yamasya Yat//

Vēdāṅga Jyōtiṣa (Yajus. verse, 42).

114. The 27 or 28 Nakṣatras are arranged in 7 groups as follows:
धृव (*or* स्थिर)–उत्तरफलगुणी, उत्तराषाढा, उत्तराभाद्रपदा, रोहिणी;
मृदु अनुराधा, चित्रा, रेवती, मृगशीर्ष;
क्षिप्र (*or* लघु)–हस्ता, अश्विनी, पुष्य, अभिजित्;
उग्र–पूर्वफल्गुणी, पूर्वाषाढा, पूर्वाभाद्रपदा, भरणी, मघा;
चर–पुनर्वसु, श्रवण, धनिष्ठा, स्वाती, शतभिषक्;
क्रूर (*or* तीक्ष्ण)–मूल, ज्येष्ठ, आर्द्रा, आश्लेषा;
मिश्रित (*or* मुदुतीक्ष्ण *or* साधारण)–कृत्तिका, विशाखा।

बृहत्संहिता, ९८. ६–११.

Dhṛva (or Sthira)—Uttaraphalguṇī, Uttarāṣāḍhā, Uttarābhādra-padā, Rōhiṇī;
Mṛdu—Anurādha, Citrā, Rēvatī, Mṛgaśīrṣa;
Kṣipra (or Laghu)—Hastā, Aśvinī, Puṣya, Abhijit;
Ugra—Pūrvaphalguṇī Pūrvāṣāḍhā, Pūrvābhādrapadā, Bharaṇī Maghā;
Cara—Punarvasu, Śravaṇa, Dhaniṣṭhā, Svātī, Śatabhiṣak;
Krūra (or Tīkṣṇa)—Mūla, Jyēṣtha, Ārdrā, Āślēṣā;
Miśrita (or Mṛdutīkṣṇa or Sādhāraṇa)—Kṛttikā, Viśākhā/

Bṛhatsaṃhitā, 98. 6-11.

115. संज्ञातुल्यमिहाचरन्ति सुधियो कार्यं हि संसिद्धये।

मुहूर्त मारतांडम्, २. ३.

Sañjñātulyamihācaranti Sudhiyō Kāryaṃ Hi Saṃsiddhayē/

Muhūrta Māratāṇḍa, 2. 3.

116. निन्द्यन्ते नित्ययोगेष्वपि नव परिघः सव्यतीपातवज्रो
व्याघातो वैधृतिश्च प्रथम परिवृताः शूलगण्डातिगण्डाः।

मुहूर्तदर्शनम्, २. १६.

Nindyantē Nityayōgēṣvapi Nava Parighaḥ Savyatīpātavajrō
Vyāghātō Vaidhṛtiśca Prathama Parivṛtāḥ Śūlagaṇḍātigaṇḍāḥ/

Muhūrtadarśana, 2. 16.

117. विरुद्ध संज्ञा इह ये च योगस्तेषामनिष्टः खलु पाद आद्यः।
सवैधृतिस्तु व्यतीपात नामा सर्वोप्यनिष्टः परिघस्य चार्धम्॥
तिस्त्रस्तु योगे प्रथमे सवज्रे व्याघातसंज्ञे नव पङ्च सूले।
गण्डातिगण्डे च षडेव नाड्यः शुभेषु कार्येषु परिवर्जनीयाः॥

रत्नमाल, ४. ४–५.

Viruddha Sañjñā Iha Yē Ca Yōgastēṣāmaniṣṭaḥ Khalu Pāda Ādyaḥ I Savaidhṛtistu Vyatīpāta Nāmā Sarvōpyaniṣṭaḥ Parighasya Cārdham II

Tisrastu Yōgē Prathamē Savajrē Vyāghātasañjñē Nava Pañca Sūlē I Gaṇḍātigaṇḍē Ca Ṣaḍēva Nāḍyaḥ Śubhēṣu Kāryēṣu Parivarjanīyāḥ II

Ratnamāla, IV. 4-5.

118. विष्कम्भे घटिकास्तिस्त्रः शूले पङ्च विवर्जयेत्।
षट्षड् गण्डेतिगण्डे च नव व्याघातवज्रयोः॥
परिघे च व्यतीपाते उभयोरपि तद्दनम्।
वैधृतौ तद्दिनं चैव यात्रायुद्धादिकं त्यजेत्॥

अग्निपुराणम्, १२७. १-२.

Viṣkambhē Ghaṭikāstisraḥ Śūlē Pañca Vivarjayēt I
Ṣaṭṣaḍ Gaṇḍētigaṇḍē Ca Nava Vyāghātavajrayōḥ II
Parighē Ca Vyatīpātē Ubhayōrapi Taddanam I
Vaidhṛtau Taddinaṃ Caiva Yātrāyuddhādikaṃ Tyajēt II

Agnipurāṇa, 127. 1-2.

119. पूर्वः स्यादुपवासादावुत्तरो व्रतदानयोः।
योगः श्राद्धे कर्मकाल व्याप्तस्तु परिगृह्यताम्॥

कालनिर्णय कारिक, १०८-१०९.

Pūrvaḥ Syādupavāsādāvuttarō Vratadānayōḥ I
Yōgaḥ Śrāddhē Karmakāla Vyāptastu Parigṛhyatām II

Kālanirṇaya Kārika, 108-109.

योगेषु तु कस्यचिल्लिङ्गस्य पूर्वेद्युर्निशीथमात्रव्याप्तिं प्रत्यसूचकत्वात्सायं-कालादिव्याप्तिर् ग्राह्या। यदि योगः पूर्वद्युर्निशीथमात्रं व्याप्नुयात् तदापरेद्युरहनि पारणं प्राप्नुयात् रात्रौ पारणस्य निषिद्धत्वात्।......तस्माद्विष्कम्भादयोगे उपवासादौ पूर्वविद्धो ग्रहीतव्यः दानव्रतयोरूदयव्यापी ग्राह्यः श्राद्धस्य तु कर्मकालव्यापी इति निर्णेतव्यम्॥

कालनिर्णय, p. ३२९-३३०.

Yōgēṣu Tu Kasyacilliṅgasya Pūrvēdyurniśīthamātra Vyāptiṃ Pratyasūca Katvāsāyaṅkālādivyāptir Grāhyā I Yadi Yōgaḥ Pūrvadyurniśīthamātraṃ Vyāpnuyāt Tadāparēdyurahani Pāraṇaṃ Prāpnuyāt Rātrau Pāraṇasya Niṣiddhatvāt.... Tasmādviṣkambhādayōgē Upavāsādau Pūrvaviddhō Grahītavyaḥ Dānavratayōrudayavyāpī Grāhyaḥ Śrāddhasya Tu Karmakālavyāpī Iti Nirṇētavyam I

Kālanirṇaya, p. 329-330.

120. History of Dharmaśāstras, Vol. IV. pp. 382-383 and note 861.

121. L.D. Swamy Kannu Pillai, Indian Chronology (Solar, Lunar & Planetary) B.C. 1 to A.D. 2000, Chapter VII, p. 22.

122. श्रवणाश्विधनिष्ठार्द्रा-नागदैवतमस्तके।
यद्यमा रविवारेण व्यतीपातः स उच्यते॥

वृद्धमनु quoted by अपरार्क, p. ४२६.
हेमाद्रि on काल, p. ६७३.

Śravaṇāśvidhaniṣṭhārdrā—Nāgadaivatamastakē I
Yadyamā Ravivārēṇa Vyatīpātaḥ Sa Uccatē II

Vṛddhamanu quoted by Aparāśrka, p. 426.
Hēmādri on Kāla, p. 673.

123. पञ्चाननस्थौ गुरु भूमि पुत्रौ मेषे रविः स्याद्यदि शुक्लपक्षे।
पाशाभिधाना करभेन युक्ता तिथिर्व्यतीपात इतीह योगः॥

स्मृत्यन्तर quoted by हेमाद्रि on काल, p. ६७३;
कालविवेक, p. ३६४.

Pañcānanasthau Guru Bhūmi Putrau Mēṣē Raviḥ Syādyadi Śuklapakṣē I
Pāśābhidhānā Karabhēna Yuktā Tithirvyatīpāta Itīha Yōgaḥ II

Smṛtyantara quoted by Hēmādri on Kāla, p. 673.
Kālavivēka, p. 364.

124. अस्मिन्हि गोभूमिहिरण्यवस्त्रदानेन सर्वं प्रविहाय पापम्।
सुरत्वमिन्द्रत्वमनामयत्वं मर्थ्याधिपत्यं लभते मनुष्यः॥

Quoted by हेमाद्रि on काल, p. ६७३-६७४.

Asminhi Gōbhūmihiraṇyavastradānēna Sarvaṃ Pravihāya Pāpam I
Suratvamindratvamanāmayatvaṃ Marthyādhipatyaṃ Labhatē Manuṣyaḥ II

Quoted by Hēmādri on Kāla, p. 673-674.

125. एकायनगतौ यातां सूर्य चंद्रमसौ यदा।
तद्युतौ मण्डले क्रान्त्योस्तुल्यत्वे वैधृता भिधः।
विपरीतायनगतौ चन्द्रार्कौ क्रान्तिलिप्तिकाः।
समास्तदा व्यतीपातो भगणार्धे तयोर्युतौ॥

सूर्य सिद्धान्त, ११. १-२.

Ēkāyanagatau Yātāṃ Sūrya Candramasau Yadā I
Tadyutau Maṇḍalē Krāntyōstulyatvē Vaidhṛtā Bhidhaḥ II
Viparītāyanagatau Candrārkau Krāntiliptikāḥ I
Samāstadā Vyatīpātō Bhagaṇārdhe Tayōryutau II

Sūrya Siddhānta, XI. 1-2.

126. तिथिं द्वि धा करोतीति करणमित्यर्थ संज्ञानात् तिथ्यर्थं करणमिति विज्ञापयते।

दीपिका on मुहूर्तदर्पणम्, १. ४१.

Tithiṃ Dvi Dhā Karōtīti Karaṇamityartha Sañjñānāt Tithyarthaṃ Karaṇamiti Vijñāpayatē I

Dīpikā on Muhūrtadarpaṇa, I. 41.

127. शेषं बवो बालवश्च कौलवस्तैतिलो गरः।
वणिजोभ्रे भवेद्विष्टिः कृष्णभूतापरार्धतः।
शकुनिर्नागश्च चतुष्पदं किंस्तुघ्नमेवच॥

नारद पुराणम्, १. ५४. १२६-१२७.

Śēṣaṃ Bavō Bālavaśca Kaulavastaitilō Garaḥ I
Vaṇijōbhrē Bhavēdviṣṭiḥ Kṛṣṇabhūtāparārdhataḥ I
Śakunirnāgaśca Catuṣpadaṃ Kiṃstughnamēvaca I I

Nārada Purāṇam, I. 54. 126-127.

128. Bṛhajjātaka, I. 9.

129. कुर्याद्बवे शुभचरस्थिर पौष्टिकानि धर्मक्रियाद्विजहितानि च बालवाख्ये।
सम्प्रीति मित्रवरणानि च कौलवे स्युः सौभाग्यसंश्रयगृहाणि च तैतिलाख्ये॥

बृहत्संहिता, ९९. ३.

Kuryādbavē Śubhacarasthirapauṣṭikāni Dharmakriyādvijahitāni Ca Bālavākhyē I
Samprītimitravaraṇāni Ca Kaulavē Syuḥ Saubhāgyasaṃśraya-gṛhāṇi Ca Taitilākhyē I I

Bṛhatsaṃhitā, 99. 3.

कृषिबीजगृहाश्रयजानि गरे वणिजि ध्रुवकार्यवणिग्युतयः।
न हि विष्टिकृतं विदधाति शुभं परिघातविषादिषु सिद्धिकरम्॥४॥

Kṛṣibījagṛhāśrayajāni Garē Vaṇiji Dhruvakāryavaṇigyutayaḥ I
Na Hi Viṣṭikṛtaṃ Vidadhāti Śubhaṃ Parighātaviṣādiṣu Siddhi-karam I I4 I I

कार्यं पौष्टिकमौषधादि शकुनौ मूलानि मन्त्रस्तथा
गोकार्याणि चतुष्पदे द्विजपितृनुद्दिश्य राज्यानि च।
नागे स्थावरदारूणानि हरणं दौर्भाग्यकर्माण्यतः
किंस्तुघ्ने शुभमिष्टिपुष्टिकरणं मङ्गल्यसिद्धि क्रियाः॥५॥

Kāryaṃ Pauṣṭikamauṣadhādi Śakunau Mūlāni Mantrastathā
Gōkāryāṇi Catuṣpadē Dvijapitṛnuddiśya Rājyāni Ca I
Nāgē Sthāvaradāruṇāni Haraṇaṃ Daurbhāgyakarmāṇyataḥ
Kiṃstughnē Śubhamiṣṭipuṣṭikaraṇaṃ Maṅgalyasiddhi Kriyāḥ I I5 I I

130. रमध्वं ने वचसे सोम्याय ऋतावरीरूप मुहूर्तमेवैः।
प्र सिन्धुमच्छा बृहती मनीषा वस्युरह्वे कुशिकस्य सूनुः॥

ऋग्वेद, ३. ३३. ५.

Ramadhvaṃ Mē Vacasē Sōmyāya Ṛtāvarīrūpa Muhūrtamēvaiḥ I

Pra Sindhumacchā Bṛhatī Manīṣā Vasyurahvē Kuśikasya Sūnuḥ //

Ṛgvēda, III. 33. 5.

रूपं रूपं मघवा बोभवीती मायाः कृण्वानस्तन्वं परि स्वाम्।
त्रिर्यद् दिवः परि मुहूर्त मागात् स्वैर्मन्त्रैरनृतुपा ऋतावा॥

ऋग्वेद, ३. ५३. ८.

Rūpaṃ Rūpaṃ Maghavā Bōbhavītī Māyāḥ Kṛṇvānastanvaṃ Pari Svām /
Triryad Divaḥ Pari Muhūrta Māgāt Svairmantrairanṛtupā Ṛtāvā //

Ṛgveda, III. 53. 8.

131. तन्मुहूर्तं धारइत्वा।

शतपथ ब्राह्मणम्, १. ८. ३. १७.

Tanmuhūrtaṃ Dhāraitvā.

Śatapatha Brāhmaṇa, I. 8. 3. 17.

अथ प्रातः अनशीत्वा मुहूर्तं सभायां आसित्वापि।

शतपथ ब्राह्मणम्, २. ३. २. ५.

Atha Prātaḥ Anaśītvā Muhūrtaṃ Sabhāyāṃ Āsitvāpi.

Śatapatha Brāhmaṇa, II. 3. 2. 5.

132. स पञ्चदशाह्नो रूपाण्य पश्यदात्मनस्तन्वो मुहूर्तांल्लोकम्पृणाः पञ्चदशैव रात्रेस्तद्यन्मुहुस्त्रायन्ते तस्मान्मुहूर्ताः।

शतपथब्राह्मणम्, १०. ४. २. १८.

Sa Pañcadaśāhnō Rūpāṇya Paśyadātmanastanvō Muhūrtāmllōkampṛṇāḥ Pañcadaśaiva Rātrēstadyanmuhustrāyantē Tasmānmuhūrtāḥ.

Śatapathabrāhmaṇa, X. 4. 2. 18.

दश च वै सहस्त्राण्यष्टौ च शतानि सम्वत्सरस्या मुहूर्ता यावन्तो मुहूर्तास्तावन्ति पञ्चदश कृत्वः क्षिप्राणि यावन्ति क्षिप्राणि तावन्ति पञ्चदशकृत्व एतर्हीणि।

शतपथब्राह्मणम्, १२. ३. २. ५.

Daśa Ca Vai Sahastrāṇyaṣṭau Ca Śatāni Samvatsarasyā Muhūrtā Yāvantō Muhūrtāstāvanti Pañcadaśa Kṛtvaḥ Kṣiprāṇi Yāvanti Kṣiprāṇi Tāvanti Pañcadaśakṛtva Ētarhīṇi /

Śatapathabrāhmaṇa, XII. 3. 2. 5.

133. चित्रः, केतूः, प्रभान्, आभान्, संभान्, ज्योतिष्मान्, तेजस्वान्, आतपन्, तपन्, अभितपन्, रोचनः, रोचमानः, शोभनः, शोभमानः कल्याणः।

तैत्तिरीय ब्राह्मणम्, ३. १०. १.

Citraḥ, Kētūḥ, Prabhān, Ābhān, Sambhān, Jyōtiṣmān, Tējasvān, Ātapan, Tapan, Abhitapan, Rōcanaḥ, Rōcamānaḥ, Śōbhanaḥ, Śōbhamānaḥ, Kalyāṇaḥ /

Taittirīya Brāhmaṇa, III. 10. 1.

All the above are the 15 Muhūrtas of the day;

The 15 Muhūrtas of the night are :
दाता, प्रदाता, आनन्दः, मोदः, प्रमेदः,
आवेशयन्, निवेशयन्, संवेशनः, संशान्तः, शान्तः,
आभवन्, प्रभवन्, संभवन्, संभूत, भूतः।
Dātā, Pradātā, Ānandaḥ, Mōdaḥ, Pramōdaḥ, Āvēśayan, Nivēśayan, Saṃvēśanaḥ, Saṃśāntaḥ, Śāntaḥ, Ābhavan, Prabhavan, Sambhavan, Sambhūtaḥ, Bhūtaḥ I

134. अथ ब्राह्मे मुहूर्ते उत्थाय काल एव प्रातरग्निहोत्रं जुहोति।

बौद्धायन गृह्यसूत्र, २. १०. २६.

Atha Brāhmē Muhūrtē Utthāya Kāla Ēva Prātaragnihōtraṃ Juhōti I

Bauddhāyana Gṛhyasūtra, II. 10. 26.

ब्राह्मे मुहूर्ते बुध्येत धर्मार्थौ चानुचिन्तयेत्।

मनुस्मृति, ४. ९२.

Brāhmē Muhūrtē Budhyēta Dharmārthau Cānucintayēt I

Manusmṛti, IV. 92.

ब्राह्मे मुहूर्ते चोत्थाय चिन्तयेदात्मनी हितम्।

याज्ञवल्क्य स्मृति, १. ११५

Brāhmē Muhūrtē Cōtthāya Ciṇtayēdātmanō Hitam /

Yājñavalkya Smṛti, I. 115.

135. Mahābhārata, Drōṇaparva, 80. 23.

136. संज्ञाः पुराणकथिता रौद्रः श्वेतस्तथा मैत्रः।
आरभटः सावित्रो वैराजस्चाथ गान्धर्वः।
अभिजिच्च रौहिणबलौ विजयाख्यो नैर्ऋतः शाक्रः।
वरूणभगदैवत्यावति विज्ञेया दिवा मुहूर्तानाम्॥
अभिजिद्वैराजश्च श्वेत सावित्र मैत्र बालविजयाः।
शुभकार्यसिद्धिजनकाः सप्त प्रोक्ताः पुराणज्ञैः॥

विध्यामाधवीय, ४. p. २७७.

Sañjñāḥ Purāṇakathitā Raudraḥ Śvētastathā Maitraḥ I
Ārabhaṭaḥ Sāvitrō Vairājascātha Gāndharvaḥ I
Abhijicca Rauhiṇabalau Vijayākhyō Nairṛtaḥ Śākraḥ I
Vāruṇabhagadaivatyāvati Vijñēyā Divā Muhūrtānām II
Abhijidvairājaśca Śvēta Sāvitra Maitra Bālavijayāḥ I
Śubhakāryasiddhijanakāḥ Sapta Prōktāḥ Purāṇajñaiḥ II

Vidhyāmādhavīya, IV. p. 277.

137. स भवान् पुष्ययोगेन मुहूर्तेन जयेन च।
कौरवेयन्प्रयात्वाशु कौन्तेयस्यार्थ सिद्धये।

महाभारतम्, उद्योगपर्वम्, ६. १७-१८.

Sa Bhavān Puṣyayōgēna Muhūrtēna Jayēna Ca/
Kauravēyanprayātvāśu Kauntēyasyārtha Siddhayē//
Mahābhārata, Udyōgaparvan, 6. 17-18.

138. रौद्रः सार्व स्तथा मैत्रः पिण्ड्य वासव एव च।
आप्योथ वैश्वदेवश्च ब्राह्मी मध्याह्नसंश्रितः।
प्राजापत्यस्तथा सेन्द्रस्तथेन्द्रो निर्ऋतिस्तथा।
वारूणश्च तथार्यम्णो भागाश्चापि दिनाश्रिताः।
वायु पुराणम्, ६६. ४०-४२.

Raudraḥ Sārva Stathā Maitraḥ Piṇḍya Vāsava Ēva Ca/
Āpyōtha Vaiśvadēvaśca Brāhmī Madhyāhnasaṃśritaḥ/
Prājāpatyastathā Aindrastathēndrō Nirṛtistathā/
Vāruṇaśca Tathāryamṇō Bhāgāścāpi Dināśritāḥ/
Vāyu Purāṇa, 66. 40-42.

अजस्तथाहिर्बुध्न्यश्च पूषाहि यमदेवताः।
आग्नेयश्चापि विज्ञेयः प्राजापत्यस्तथैव च।
ब्रह्मा सौम्यस्तथादित्यो बार्हस्पत्योथ वैष्णवः।
सावित्रोथ तथा त्वष्ट्रो वायव्यश्चेति संग्रहः।
एकरात्रिमुहूर्ताः स्युः क्रमोक्त दश पञ्च च।
वायु पुराणम्, ६६. ४३-४४.

Ajastathāhirbudhnyaśca Pūṣāhi Yamadēvatāḥ/
Āgnēyaścāpi Vijñēyaḥ Prājāpatyastathaiva Ca/
Brahmā Saumyastathādityō Bārhaspatyōtha Vaiṣṇavaḥ/
Sāvitrōtha Tathā Tvaṣṭrō Vāyavyaścēti Saṅgrahaḥ/
Ēkarātrimuhūtāḥ Syuḥ Kramōkta Daśa Pañca Ca/
Vāyu Purāṇa, 66. 43-44.

139. विष्कंभगण्डपरिघ वज्रयोगेषु कारायेत्।
स्वेते मैत्रेथ माहेन्द्रे गान्धर्वभिजिति रौहिणे॥
तथा वैराज सावित्रे मुहूर्ते गृहमरभेत्।
चन्द्रादित्यबलं लब्ध्वा शुभलग्नं निरीक्षयेत्॥
मत्स्य पुराणम्, २५३. ८-९.

Viṣkambhagaṇḍaparigha Vajrayōgēṣu Kārāyēt/
Svētē Maitrētha Māhēndrē Gāndharvabhijiti Rauhiṇē//
Tathā Vairāja Sāvitrē Muhūrtē Gṛhamarabhēt/
Candrādityabalaṃ Labdhvā Śubhalagnaṃ Nirīkṣayēt//
Matsya Purāṇa, 253. 8-9.

140. यत्कार्यं नक्षत्रे तद्देवत्यासु तिथिषु तत्कार्यम्।
करणमुहूर्त्तेष्वपि तत् सिद्धिकरं देवतासदृशम्॥
बृहत्संहिता, ९९. ३.
Yatkāryaṃ Nakṣatrē Taddēvatyāsu Tithiṣu Tatkāryam।
Karaṇamuhūrttēṣvapi Tat Siddhikaraṃ Dēvatāsadṛśam॥
Bṛhatsaṃhitā, 99 3.

141. कालः शुभक्रियायोग्यो मुहूर्त इति कथ्यते।
मुहूर्तदर्शन of विध्यामाधवीय, १. २०.
Kālaḥ Śubhakriyāyōgyō Muhūrta Iti Kathyate।
Muhūrtadarśana of Vidhyāmādhavīya, I. 20.

142. उदगयने सितपक्षे शिशिरगभस्तौ च जीववर्गस्थे।
लग्ने स्थिरे स्थिरांशे सौम्यैर्धीधर्मकेन्द्रगतैः॥
पापैरूपचयसंस्थैर्ध्रुवमृदुहरितिष्यवायुदेवेषु।
विकुजे दिनेनुकूले देवानां स्थापनं शस्तम्॥
बृहत्संहिता, ६०. २०-२१.
Udagayanē Sitapakṣē Śiśiragabhastau Ca Jīvavargasthē।
Lagnē Sthirē Sthirāṃśē Saumyairhīdharmakēndragataiḥ॥
Pāpairūpacayasaṃsthairdhruvamṛduharitiṣyavāyudēvēṣu।
Vikujē Dinēnukūlē Dēvānāṃ Sthāpanaṃ Śastam॥
Bṛhatsaṃhitā, 60. 20-21.

143. E.I., Vol. VI. p. 332-349, at p. 340.

144. यत् त्वा सूर्य स्वर्भानुस्तमसाविध्यदासुरः।
अक्षेत्रविद् यथा मुग्धो भुवनान्यदीधयुः॥
स्वर्भनोरधयदिन्द्र माया अवो दिवो वर्तमाना अवाहन्।
गूळ्हं सूर्यं तमसापव्रतेन तुरीयेण ब्रह्मणाविन्ददत्रिः॥
ग्राव्णो ब्रह्मा युयुजानः सपर्यन् कीरिणा देवान् नामसोपशिक्षन्।
अत्रिः सूर्यस्य दिवि चक्षुराधात् स्वर्भानोरप माया अघुक्षत्॥
ऋग्वेद, ५. ४०. ५, ६, ८.
Yat Tvā Sūrya Svarbhānustamasāvidhyadāsuraḥ।
Akṣētravid Yathā Mugdhō Bhuvanānyadīdhayuḥ॥
Svarbhanōradhayadindra Māyā Avō Divō Vartamānā Avāhan।
Gūlhaṃ Sūryaṃ Tamasāpavratēna Turīyēṇa Brahmaṇāvindadatriḥ॥
Gravṇō Brahmā yuyujānaḥ Saparyan Kīrinā Dēvān Nāmasōpaśikṣan।
Atriḥ Sūryasya Divi Cakṣurādhāt Svarbhānōrapa Māyā Aghukṣat॥

Ṛgvēda, V. 40. 5. 6. 8.

145. स्वर्भानुर्हासुर आदित्यं तमसाविध्यत्तस्यात्रय स्तमौप जिघांसन्त एतं सप्त दशस्तोमं त्र्यहं पुरस्ताद्विषुवत् उपायंस्तस्य परस्तात्तमोपजघ्नुः।

सांख्यायण ब्राह्मणम्, २४. ३.

Svarbhānurhāsura Ādityaṃ Tamasāvidhyattasyātraya Stamaupa Jighāṃsanta Ētaṃ Sapta Daśastōmaṃ Tryahaṃ Purastādviṣuvat Upāyaṃstasya Parastāttamōpajaghnuḥ I

Sāṅkhyāyaṇa Brāhmaṇa, 24. 3.

146. भूच्छायां स्वग्रहणे भास्करमर्कग्रहे प्रविशतीन्दुः।
प्रग्रहणमतः पश्चान्नेन्दोर्भानोश्च पूर्वार्द्धात्॥
एवमुपरागकारणमुक्तमिदं दिव्यदृग्भिराचार्यैः।
राहुरकारणमस्मिन्नित्युक्तः शास्त्रसद्भावः॥

बृहत्संहिता, ५. ८ & १३.

Bhūcchāyāṃ Svagrahaṇē Bhāskaramarkagrahē Praviśatīnduḥ I
Pragrahaṇamataḥ Paścānnēndōrbhānōśca Pūrvārddhāt II
Ēvamuparāgakāraṇamuktamidaṃ Divyadṛgbhirācāryaiḥ I
Rāhurakāraṇamasminnityutkaḥ Śāstrasadbhāvaḥ I

Bṛhatsaṃhitā, 5. 8 & 13.

147. व्यासः। इन्दोर्लक्षगुणं प्रोक्तं रवेर्दशगुणं स्मृतम्।
गङ्गातोय इन्दोः कोटी रवेर्दश॥

हेमाद्रि on काल, ३८४; कालविवेक, ५२१; निर्णयसिन्धु, ६४.

Vyāsaḥ. Indōrlakṣaguṇaṃ Prōktaṃ Ravērdaśaguṇaṃ Smṛtam/
Gaṅgātōya Indōḥ Kōṭī Ravērdaśa//

Hēmādri on Kāla, 384; Kālavivēka, 521; Nirṇayasindhu, 64.

148. सर्वेषामेव वर्णानां सूतकं राहुदर्शने।
स्नात्वा कर्मणि कुर्वीत शृतमन्नं विवर्जयेत्॥

षट्त्रिंशन्मत, quoted by कालविवेक, ५३३.

Sarvēṣāmēva Varṇānāṃ Sūtakaṃ Rāhudarśanē I
Snātvā Karmaṇi Kurvīta Śṛtamannaṃ Vivarjayēt II

Ṣaṭtriṃśanmata, quoted by Kālavivēka, 533.

सर्वेषा...राहुसूतके।
सचेलं तु भवेत्स्नानं सूतकान्नं च वर्जयेत्॥
अयने विषुवे चैव चन्द्रसूर्यग्रहे तथा।
अहोरात्रोषितः स्नात्वा सर्व पापैः प्रमुच्यते॥

विष्णु पुराणम् quoted by हेमाद्रि on काल, ३८३.

Sarvēṣā...Rāhusūtakē I
Sacēlaṃ Tu Bhavētsnānaṃ Sūtakānnaṃ Ca Varjayēt II

Ayanē Viṣuvē Caiva Candrasūryagrahē Tathā I
Ahōrātrōṣitaḥ Snātvā Sarva Pāpaiḥ Pramucyatē II
Viṣṇu Purāṇa, quoted by Hēmādri on Kāla, 383.

149. ग्रहण सन्क्रमे चैव न स्नायाद्यस्तु मानवः।
सप्तजन्मसु कुष्ठी स्याद् दुःख भागी न संशयः।
समयमयूख, १३०.

Grahaṇa Sankramē Caiva Na Snāyādyastu Mānavaḥ I.
Saptajanmasu Kuṣṭhī Syād Duḥkha Bhāgī Na Saṃśayaḥ I
Samayamayūkha, 130.

चन्द्रे वा यदि वा सूर्ये दृष्टे राहौ महाग्रहे।
अक्षयं कथितं पुण्यं तत्रार्के तु विशेषतः॥
मार्कण्डेय पुराणम्, quoted by हेमाद्रि on काल, ३८४.

Candrē Vā yadi Vā Sūryē Dṛṣṭē Rāhau Mahāgrahē I
Akṣayaṃ Kathitaṃ Puṇyaṃ Tatrārkē Tu Viśēṣataḥ II
Mārkaṇḍēya Purāṇa, quoted by Hēmādri on Kāla, 384.

150. सर्वं गङ्गासमं तोयं सर्वे व्याससमा द्विजाः।
सर्वं मेरूसमं दानं ग्रहणे सूर्यचंद्रयोः।
भुजबलनिबन्ध of भोज, p. ३४८; कालनिर्णय, ३४८; समयमयूख, १३०.

Sarvaṃ Gaṅgāsamaṃ Tōyaṃ Sarvē Vyāsasamā Dvijāḥ I
Sarvaṃ Mērūsamaṃ Dānaṃ Grahaṇē Sūryacandrayōḥ II
Bhujabalanibandha of Bhōja, p. 348; Kālanirṇaya, 348; Samayamayūkha, 130.

गोदावरी भिमरथी तुङ्गभद्र च वेणिका।
तापी पयोश्णी विन्ध्यस्य दक्षिणे तु प्रकीर्तिताः॥
भागीरथी नर्मदा यमुना च सरस्वती।
विशोका च वितस्ता च हिमवत्पर्वताश्रिताः॥
एता नद्यः पुण्यतमा देव तीर्थान्युदाहृताः।
ब्रह्मपुराणम्, ७०. ३३-३५.

Gōdāvarī Bhimarathī Tuṅgabhadra Ca Vēṇikā I
Tāpī Payōṣṇī Vindhyasya Dakṣiṇē Tu Prakīrtitāḥ II
Bhāgīrathī Narmadā Yamunā Ca Sarasvatī I
Viśōkā Ca Vitastā Ca Himavatparvatāśritāḥ II
Ētā Nadyaḥ Puṇyatamā Dēva Tīrthānyudāhṛtāḥ I
Brahmapurāṇa, 70. 30-35.

151. ग्रस्यमाने भवेत्स्नानं ग्रस्ते होमो विधीयते।
मुच्यमाने भवेद्दानं मुक्ते स्नानं विधीयते॥
भुजबल, p. ३४७. Verse १५१३, quoted by कालनिर्णय, ३५३; हेमाद्रि on काल, ३९०.

Grasyamānē Bhavētsnānaṃ Grastē Hōmō Vidhīyatē /
Mucyamānē Bhavēddānaṃ Muktē Snānaṃ Vidhīyatē //
Bhujabala, p. 347, verse 1513, quoted by Kālanirṇaya, 353; Hēmādri on Kāla, 390.
ग्रहस्पर्श काले स्नानं मध्ये होमः सुरार्चनं श्राद्धं च।
मुच्यमाने दानं मुक्ते स्नानमिति क्रमः।
...एवं जप होम तर्पण मार्जन विप्र भोजनात्मक पञ्चप्रकारं
पुरश्चरणं तर्पणाद्य सम्भवे तत्तत्संख्या चतुर्गुणो जप एत कार्यः॥

धर्मसिन्धु, ३२-३५.

Grahasparśa Kālē Snānaṃ Madhyē Hōmaḥ Surācanaṃ Śrāddhaṃ Ca /
Mucyamānē Dānaṃ Muktē Snānamiti Kramaḥ /
...Ēvaṃ/ Japahōmatarpaṇamārjanaviprabhōjanātmaka Pañcaprakāraṃ Puraścaraṇaṃ Tarpaṇādya Sambhavē Tattatsaṅkhyā Caturguṇō Japa Ēta Kāryaḥ //

Dharmasindhu, 32-35.

152. **तत्र जननमरण शौचिनामपि स्नानं**
कर्तव्यं दानं श्राद्धं च न कर्तव्यम्।

कृत्यतत्त्व, p. ४३३.

Tatra Jananamaraṇa Śaucināmapi Snānaṃ Kartavyaṃ Dānaṃ *Śrāddhaṃ Ca Na Kartavyam* /

Kṛtyatattva, p. 433.

153. **अत्र शौचमध्येपि स्नानश्राद्धादि कार्यमेव। सूतके मृतके चैव न दोषो राहुदर्शने।**
तावदेव भवेच्छुद्धिर्यावन्मुक्तिर्न दृश्यते इति माधवीये वृद्धवसिष्ठोक्ते। ...इयं च
शुद्धिरवि शेषान्मन्त्रदीक्षापुरश्चरणादि सर्व स्मार्त कर्मविषया। मदनरत्नेप्येवम्।

निर्णयसिन्धु, ६६.

Atra Śaucamadhyēpi Snānaśrāddhādi Kāryamēva /
Sūtakē Mṛtakē Caiva Na Dōṣō Rāhudarśanē //
Tāvadēva Bhavēcchuddhiryāvanmuktirna Dṛśyatē Iti Mādhavīyē Vṛddha Vasiṣṭhōktē /
...Iyaṃ Ca Śuddhiravi Śēṣānmantradīkṣā Puraścaraṇādi Sarva Smārta Karmaviṣayā / *Madanaratnēpyēvam* /

Nirṇayasindhu, 66.

154. **कार्तिके ग्रहणं श्रेष्ठं गङ्गायमुनसङ्गमे।**
मार्गे तु ग्रहणं प्रोक्तं देविकायां महामुने॥
पौषे तु नर्मदा पुण्या माघे सन्निहिता शुभा।
देवी पुराणम् quoted by कृत्यकल्प (नैयतकाल) ३७०-७१; हेमाद्रि on काल, ३८५-८६; कालविवेक, ५२४.

Kārtikē Grahaṇaṃ Śrēṣṭhaṃ Gaṅgāyamunasaṅgamē I
Mārgē Tu Grahaṇaṃ Prōktaṃ Dēvikāyāṃ Mahāmunē II
Pauṣē Tu Narmadā Puṇyā Māghē Sannihitā Śubhā I
Dēvī Purāṇa quoted by Kṛtyakalpa (Naiyatakāla), 370-71; Hēmādri on Kāla, 385-86; Kālavivēka, 524.

155. सशिरावमज्जनमप्सु वर्जयेत। अस्तमिते च स्नानं।

आपस्तंब धर्मसूत्र, १. ११. ३२. ८.

Saśirāvamajjanamapsu Varjayēta I Astamitē Ca Snānaṃ I

Āpastamba Dharmasūtra, I. 11. 32. 8.

156. रात्रौ श्राद्धं न कुर्वति राक्षसी कीर्तिता हि सा।
सन्ध्ययोरूभयोश्चैव सूर्ये चैवाचिरोदिते॥

मनुस्मृति, ३. २८०.

Rātrau Śrāddhaṃ Na Kurvati Rākṣasī Kīrtitā Hi Sā I
Sandhyayōrūbhayōścaiva Sūryē Caivācirōditē II

Manusmṛti, III. 280.

157. व्यतीपातो गजच्छाया ग्रहणं चन्द्रसूर्ययो:।
श्राद्धं प्रति रुचिश्चैते श्राद्ध काल: प्रकीर्तिता॥

याज्ञवल्क्य स्मृति, १. २१८.

Vyatīpātō Gajacchāyā Grahaṇaṃ Candrasūryayōḥ I
Śrāddhaṃ Prati Ruciścaitē Śrāddha Kālaḥ Prakīrtitā II

Yājñavalkya Smṛti, I. 218.

158. शतातप: स्नानं दानं तप: श्राद्धमनन्तं राहुदर्शने।
आसुरी रात्रिरन्यत्र तस्मत्तां परिवर्जयेत्॥

हेमाद्रि on काल, p. ३८७; कालविवेक, ५२७; स्मृति कौस्तुभ, ७१.

Śatātapaḥ Snānaṃ Dānaṃ Tapaḥ Śrāddhamanantaṃ Rāhudarśanē I
Āsurī Rātriranyatra Tasmattāṃ Parivarjayēt II
Hēmādri on Kāla, p. 387; Kālavivēka, 527; Smṛti Kaustuba, 71.

159. महाभारते। अयने विषुवेचैव ग्रहणे चन्द्रसूर्ययो:।
पात्र भूताय विप्राय भूमिं दद्यात्सदक्षिणाम्॥

कालनिर्णय, p.३५४; स्मृति कौस्तुभ, ७२.

Mahābhāratē I Ayanē Viṣuvēcaiva Grahaṇē Candrasūryayōḥ I
Pātra Bhūtāya Viprāya Bhūmiṃ Dadyātsadakṣiṇām II

Kālanirṇaya, p. 354; Smṛti Kaustubha, 72.

160. चन्द्रसूर्यग्रहे भुक्त्वा प्राजापत्येन शुध्यति।

षट्त्रिंशन्मत, quoted by हेमाद्रि on काल, p. ३८२.

Candrasūryagrahē Bhuktvā Prājāpatyēna Śudhyati I

Ṣaṭtriṃśanmata, quoted by Hēmādri on Kāla, p . 382.

161. यद्यपि चन्द्रसूर्यग्रहे नाद्यात् इति ग्रहणे भोजननिषेध
स्तथापि भोक्तुर्दोषः दातुरभ्युदयः।

मिताक्षर onयाज्ञवल्क्य स्मृति, १. २१७-२१८.

Yadyapi Candrasūryagrahē Nādyāt Iti Grahaṇē Bhōjananiṣēdha Stathāpi Bhōkturdōṣaḥ Dāturabhyudayaḥ I

Mitākṣara on Yājñavalkya Smṛti, I. 217-218.

162. सूर्येन्द्वोर्ग्रहणं यावत्तावत्कुर्याज्जपादिकम्।
न स्नायान्न च भुञ्जीत स्नात्वा भुञ्जीत मुक्तयोः॥

शिवरहस्य, quoted by हेमाद्रि on काल, p. ३८९.

Sūryēndvōrgrahaṇaṃ Yāvattāvatkuryājjapādikam I
Na Snāyānna Ca Bhuñjīta Snātvā Bhuñjīta Muktayōḥ II

Śivarahasya, quoted by Hēmādri on Kāla, p. 389.

ग्रहणादि सप्तदिन पर्यन्तं रामगोपालाद्यागमदीक्षोक्ता शिवार्चन चन्द्रिकायां ज्ञानार्णवेमन्त्राद्यारंभणं कुर्याद् ग्रहणे चन्द्रसूर्ययोः। ग्रहणाद्वापि देवेशि कालः सप्तदिनावधि॥ इति...अत्रसूर्यग्रहणमेव मुख्यम्।

निर्णयसिन्धु, ६७.

Grahaṇādi Saptadina Paryantaṃ Rāmagōpālādyāgamadīkṣōktā Śivārcana Candrikāyāṃ Jñānānṇavē Mantrādyārambhaṇaṃ Kuryād Grahaṇē Candra Sūryayōḥ I *Grahaṇādvāpi Dēvēśi Kālaḥ Saptadināvadhi* II

Iti...Atrasūryagrahaṇamēva Mukhyam I

Nirṇayasindhu, 67.

163. संक्रान्तौ पुण्यकालस्तु षोडशोभयतः कालाः।
चन्द्रसूर्योपरागे तु यावद्दर्शनगोचरः॥

जाबालि, quoted by कृत्यकल्प (नैयत), ३६८; हेमाद्रि on काल, ३८८; कृत्यरत्नाकर, ६२५.

Saṅkrāntau Puṇyakālastu Ṣōḍaśōbhayataḥ Kālāḥ I
Candrasūryōparāgē Tu Yāvaddarśanagōcaraḥ II

Jābāli, quoted by Kṛtyakalpa (Naiyata), 368; Hēmādri on Kāla, 388; Kṛtyaratnākara, 625.

164. सूर्यग्रहः सूर्यवारे सोमे सोमग्रहस्तथा।
चूडामणिरयं योगस्तत्रानन्त फलं स्मृतं
वारेष्वन्येषु यत्पुण्यं ग्रहणे चंद्रसूर्ययोः।
तत्पुण्यं कोटिगुणितं ग्रासे चूडामणौ स्मृतम्॥

कालविवेक, ५२३; कालनिर्णय, ३५१; तिथि तत्त्व, १५४.

Sūryagrahaḥ Sūryavārē Sōmē Sōmagrahastathā I
Cūḍāmaṇirayaṃ Yōgastatrānanta Phalaṃ Smṛtaṃ
Vārēṣvanyēṣu Yatpuṇyaṃ Grahaṇē Candrasūryayōḥ I
Tatpuṇyaṃ Kōṭiguṇitaṃ Grāsē Cūḍāmaṇau Smṛtam II

Kālavivēka, 523; Kālanirṇaya, 351; Tithi Tattva, 154.

165. आदित्यग्रहणे राम ग्रहणे च निशाभृताम्।
उपवासादवाप्नोति सर्वकल्मष नाशनम्।।

हेमाद्रि on व्रत, २. ९१७. (from विष्णुधर्मोत्तर).

Ādityagrahaṇē Rāma Grahaṇē Ca Niśābhṛtām I
Upavāsādavāmōti Sarvakalmaṣa Nāśanam II

Hēmādri on Vrata, II. 917. (from Viṣṇudharmōttara).

166. तथा च जैमिनिः आदित्येहनि संक्रान्तौ ग्रहणे सूर्यचन्द्रयोः।
पारणं चोपवासं च न कुर्यात् पुत्रवान् गृही।।

हेमाद्रि on काल, p. ३८२.

Tathā Ca Jaiminiḥ Ādityēhani Saṅkrāntau Grahaṇē Sūryacandrayōḥ/
Pāraṇaṃ Cōpavāsaṃ Ca Na Kuryāt Putravān Gṛhī//

Hēmādri on Kāla, p. 382.

167. वृद्धगौतमः।।
सूर्यग्रहे तु नाशनीयात् पूर्वं यामचतुष्टयम्।
चन्द्रग्रहे तु यामांस्त्रीन् वालवृद्धातुरैर्विना।।

हेमाद्रि on काल, p. ३८१.

Vṛddhagautamaḥ II
Sūryagrahē Tu Nāśanīyāt Pūrvaṃ Yāmacatuṣṭayam I
Candragrahē Tu Yāmāṃstrīn Vālavṛddhāturairvinā II

Hēmādri on Kāla, p. 381.

168. एकस्मिन्यदि मासे स्याद् ग्रहणं चंद्रसूर्ययोः ब्रह्मक्षत्रविरोधाय विपरीते विवृद्धये।।

विष्णुधर्मोत्तर पुराणम्, १. ८५. ५६.

Ēkasminyadi Māsē Syād Grahaṇaṃ Candrasūryayōḥ Brahma-kṣatravirōdhāya Viparītē Vivṛddhayē II

Viṣṇudharmōttara Purāṇa, I. 85. 56.

169. रवेः संक्रमणं रासौ संक्रान्तिरिति कथ्यते।
स्नानदानतपः श्राद्धहोमादिषु महाफला।।

नागरखण्ड, quoted by हेमाद्रि on काल, p. ४१०.

Ravēḥ Saṅkramaṇaṃ Rāsau Saṅkrāntiriti Kathyatē I
Snānadānatapaḥ Śrāddhahōmādiṣu Mahāphalā II

Nāgarakhaṇḍa, quoted by Hēmādri on Kāla, p. 410.

170. On the day previous a person, whether male or female, should take food only once in the noon and on the Saṅkrānti day should brush his teeth and then take a bath with water mixed with sesame. The person should offer a self-restrained Brāhmaṇa householder three vessels containing edibles together with a cow intending them to be for Yama, Rudra and Dharma and repeat the verses. Further, if able should make gifts of ornaments, a bed stread and golden jars (two) to the Brāhmaṇa; but if poor should offer only fruits to the Brāhmaṇa. Then he should take his meal without oil and should feed others according to his ability.

171. न केवलमादित्यस्यैव संक्रमायने पुण्यकालः किं तु सर्वेषामपि ग्रहाणां नक्षत्रराशि संक्रमे पुण्यकालो भवति।

कालनिर्णय, p. ३४५.

Na Kēvalamādityasyaiva Saṅkramāyanē Puṇyakālaḥ Kiṃ Tu Sarvēṣāmapi Grahāṇāṃ Nakṣatrarāśi Saṅkramē Puṇyakālō Bhavati.

Kālanirṇaya, p. 345.

172. मेष तुलादौ विषुवत् षडशीतिमुखं तुलादि भागेषु।
षडशीति मुखेषु रवेः पितृदिवसा येवशेषाः स्युः।।
षडशीतिमुखं कन्या चतुर्दशेष्टादशे च मिथुनस्य।
मीनस्य द्वाविंशे षड्विंशे कार्मुकस्यांशे।।
तुला आदिर्यस्याः सा तुलदिः कन्या।
द्वादशैव भवन्त्येषां द्विज नामानि मे शृणु।
एकं विष्णुपदं नाम षडशीतिमुखं तथा।।
विषुवं च तृतीयं च अन्ये द्वे दक्षिणोत्तरे।
कुंभाव्लिगोहरिषु विष्णुपदं वहन्ति स्त्रीचापमीनमिथुने षडशीतिवक्त्रम्।
अर्कस्य सौम्यमयनं शशिधाम्नि याम्यमृक्षे झषे विषुवति त्वजतौव्लिनोः स्यात्।।

ब्रह्मवैवर्तपुराणम्, quoted by हेमाद्रि on काल, p. ४०७.

Mēṣa Tulādau Viṣuvat Ṣaḍaśītimukhaṃ Tulādi Bhāgēṣu।
Ṣaḍaśīti Mukhēṣu Ravēḥ Pitṛdivasā Ēvaśēṣāḥ Syuḥ।।
Ṣaḍaśītimukhaṃ Kanyā Caturdaśēṣṭādaśē Ca Mithunasya।
Mīnasya Dvāviṃśē Ṣaḍviṃśē Kārmukasyāṃśē।।
Tulā Ādiryasyāḥ Sā Tuladiḥ Kanyā।
Dvādaśaiva Bhavantyēṣāṃ Dvija Nāmāni Mē Śṛṇu।
Ēkaṃ Viṣṇupadaṃ Nāma Ṣaḍaśītimukhaṃ Tathā।।
Viṣuvaṃ Ca Tṛtīyaṃ Ca Anye Dvē Dakṣiṇōttarē।
Kumbhāḷigōhariṣu Viṣṇupadaṃ Vahanti Strīcāpamīnamithunē Ṣaḍaśītivaktram।

Arkasya Saumyamayanaṃ Śaśidhāmni Yāmyamṛkṣē Jhaṣē Viṣuvati Tvajatauḷinōḥ Syāt//

Brahmavaivartapurāṇa, quoted by Hēmādri on Kāla, p. 407.

173. समायनऋतुमासपक्षाहादि क्रमेण तु।
स्थूल सूक्ष्मविभागेन देवी सर्वगतविभो।
पुण्यपापविभागेन फलं देवी प्रयच्छति।
एकधापि कृतं तस्मिन्कोटिकोटिगुणं भवेत्।
धर्माद्विवर्धते ह्यायू राज्यं पुत्रसुखादयः।
अधर्माद् व्याधिशोकादि विषुवायनसंनिधौ।
विषुवेषु च यज्जप्तं दत्तं भवति चाक्षयम्।
एवं विष्णुपदे चैव षडशीतिमुखेषु च॥

देवीपुराणम्, quoted by हेमाद्रि on काल, p. ४१८-४१९.

Samāyanaṛtumāsapakṣāhādi Kramēṇa Tu/
Sthūla Sūkṣmavibhāgēna Dēvī Sarvagatavibhō/
Puṇyapāpavibhāgēna Phalaṃ Dēvī Prayacchati/
Ēkadhāpi Kṛtaṃ Tasminkōṭikōṭiguṇaṃ Bhavēt/
Dharmādvivardhatē Hyāyū Rājyaṃ Putrasukhādayaḥ/
Adharmād Vyādhiśōkādi Viṣuvāyanasannidhau/
Viṣuvēṣu Ca Yajjaptaṃ Datttaṃ Bhavati Cākṣayam/
Ēvaṃ Viṣṇupadē Caiva Ṣaḍaśītimukhēṣu Ca//

Dēvīpurāṇa, quoted by Hēmādri on Kāla, p. 418-419.

174. त्रिचतुः पञ्चसप्ताष्टनवद्वादश एव च क्रमेण घटिका ह्योतास्तत्पुण्यं पारमार्थिकम्॥

देवीपुराणम् quoted by हेमाद्रि on काल, p. ४०९.

Tricatuḥ Pañcasaptāṣṭanavadvādaśa Ēva Ca Kramēṇa Ghaṭikā Hyōtāstatpuṇyaṃ Pāramārthikam//

Dēvīpurāṇa quoted by Hēmādri on Kāla, p. 409.

175. सप्तधा सा तु बोद्धव्या एकैकैव यथा शृणु।
मन्दा मन्दाकिनी ध्वाङ्क्षी घोरा चैव महोदरी।
राक्षसी मिश्रिता प्रोक्ता संक्रान्तिः सप्तधा नृप।
सूर्ये घोरा विधौ ध्वाङ्क्षी भौमवारे महोदरी।
बुधे मन्दाकिनी नाम मन्दा सुरपुरोहिते।
मिश्रिता शुक्रवारे स्याद्राक्षसी च शनैश्चरे।

देवीपुराणम्, quoted by हेमाद्रि on काल, p. ४०८-४०९.

Saptadhā Sā Tu Bōddhavyā Ēkaikaiva Yathā Śṛṇu/
Mandā Mandākinī Dhvāṅkṣī Ghōrā Caiva Mahōdarī/

Rākṣasī Miśritā Prōktā Saṅkrāntiḥ Saptadhā Nṛpa l
Sūryē Ghōrā Vidhau Dhvāṅkṣī Bhaumavārē Mahōdarī l
Budhē Mandākinī Nāma Mandā Surapurōhitē l
Miśritā Śukravārē Syādrākṣasī Ca Śanaiścarē l

Dēvīpurāṇa quoted by Hēmādri on Kāla, p. 408-409.

176. त्रीण्युत्तराणि तेभ्यो रोहिण्यश्च ध्रुवाणि तैः कुर्यात्।
अभिषेकशान्तितरूनगरधर्मबीजध्रुवारम्भान्॥ बृहत्संहिता, ९८. ६.
Trīṇyuttarāṇi Tēbhyō Rōhiṇyaśca Dhruvāṇi Taiḥ Kuryāt l
Abhiṣēkaśāntitarunagaradharmabījadhruvārambhān l l

Bṛhatsaṃhitā, 98. 6.

मूलशिवशक्रभुजगाधिपानि तीक्ष्णानि तेषु सिध्यन्ति।
अभिघातमन्त्रवेतालबन्धभेदसम्बन्धाः॥७॥
Mūlaśivaśakrabhujagādhipāni Tīkṣṇāni Tēṣu Sidhyanti l
Abhighātamantravētālabandhabhēdasambandhāḥ l l7 l l
उग्राणि पूर्व भरणी पित्र्याण्युत्सादनाशशाठचेषु।
योज्यानि बन्धविषदहनशस्त्रघातादिषु च सिद्ध्यै॥८॥
Ugrāṇi Pūrvabharaṇīpitryāṇyutsādanāśāśāṭhacēṣu l
Yōjyāni Bandhaviṣadahanaśastraghātādiṣu Ca Siddhyai l l8 l l
लघु हस्ताश्विनपुष्याः पुण्यरतिज्ञानभूषणकलासु।
शिल्पौषधयानादिषु सिद्धिकराणि प्रदिष्टानि॥९॥
Laghu Hastāśvinapuṣyāḥ Puṇyaratijñānabhūṣaṇakalāsu l
Śilpauṣadhayānādiṣu Siddhikarāṇi Pradiṣṭāni l l9 l l
मृदुवर्गोनुराधाचित्रापौष्णैन्दवानि मित्रार्थे।
सुरतविधिवस्त्रभूषणमङ्गलगीतेषु च हितानि॥१०॥
Mṛduvargōōnurādhācitrāpauṣṇaindavāni Mitrārthē l
Suratavidhivastrabhūṣaṇamaṅgalagītēṣu Ca Hitāni l l10 l l
हौतभुजं सविशाखं मृदुतीक्ष्णं तद्धिमिश्रफलकारि।
श्रवणत्रयमादित्यानिले च चरकर्मणि हितानि॥११॥
Hautabhujaṃ Saviśākhaṃ Mṛdutīkṣṇaṃ Taddhimiṣraphalakāri l
Śravaṇatrayamādityānilē Ca Carakarmaṇi Hitāni l l11 l l

177. Hēmādri on Kāla, p. 415-416.

TABULAR STATEMENTS

Tabular Statements Showing Significance of Applications of Time In Pre-Kaliyuga and Kaliyuga Periods (Eras and Regnal Years) From Epigraphical Records

Eras

Table 1 : The Vikrama Era or Kṛta Era (58 A.D.)

Table 2 : The Śaka Era (78 A.D.)

Table 3 : The Cēdi/Kālacuri Era (249 A.D.)

Table 4 : The Gupta Era (319 A.D.)

Table 5 : The Vallabhi Era (318-319 A.D.)

Table 6 : The Harṣa Era (606-607 A.D.)

Regnal Years

Table 7 : The Mauryas (320 B.C. 185 B.C.)

Table 8 : The Śātavāhanas (75 B.C.-250 A.D.)

Table 9 : The Ikṣvākus (200 A.D.-260 A.D.)

The Vikrama Era or Kṛta Era (58 A.D.)

Sl. No.	Samvatsara	Ayana	Season/ Ṛtu	Month/ Māsa	Fortnight/ Pakṣa	Day/ Tithi	Star/ Nakṣatra	Karaṇa	Yōga	Mahūrta	Donor
1	2	3	4	5	6	7	8	9	10	11	12
1.	134	-	-	Śrāvaṇa	-	23	-	-	-	-	Canadrabhi, the d/o the householder Dharma, the w/o Bhadrapāla
2.	168	-	-	Jyēṣṭha	Bright Half/śu	15	-	-	-	-	Trapa, a lady
3.	384	-	-	Caitra	Brigh Half/śu	15	-	-	-	-	-
4.	335	-	-	Jēṣṭha/ Jyeṣṭha	Bright Half/śu	15	-	-	-	-	-
5.	359	-	-	Āśvayuja		1	-	-	-	-	By a lady hearer, Pōtaka

Donee	Object of the Inscription	Purpose of the Inscription	Corres Christian Era	Remarks or References
13	14	15	16	17
-	To record the putting up the relics in the hores Stūpa by a worshipper	Is having venerated the country town for the veneration of all beings; may it be for obtainment of Nirvāṇa	-	Kalāwan copper plate Insc. of the yr. 134; EI XXI, No. 39, p. 251.
-	To record the excavation of a well inside a certain monastery	To provide for drinking water specially during the hottest part of the year	110 A.D.	Peṣāwar Musem Insc. of the yr. 168; EI XIX, No. 31, p. 202.
-	-	-	228 A.D.	Two Yūpa Inscs. from Barnāla, Kṛta years 284 and 335; EI XXVI, No. 12, p. 119.
-	-	-	279 A.D.	Two Yūpa Inscs. from Barnāla, Kṛta years 284 and 335; EI XIX, No. 32, p. 203.
To all living beings	A jewel has been installed	-	301 A.D.	A Kharōṣṭhi Insc. from Jamalgarhi of the year 359; EI XIX No. 32, p. 203.

1	2	3	4	5	6	7	8	9	10	11	12
6.	461 MV Era	-	-	Āśvayuja	BrightHalf/śu	5	-	-	-	-	-
7.	524 M Sam	-	Spring	-	-	-	-	-	-	-	Dattabhaṭṭa, the commander of the forces of King Prabhākar
8.	547	-	-	Māgha	Bright half/ Śuddha	10	-	-	-	-	-
9.	611	-	Autumn	-	-	-	-	-	-	-	Sūryavarman, the accomplished s/o Maukari King Īśānavarman
10.	682	-	-	-	-	-	-	-	-	-	-
11.	699	-	-	Āṣāḍha	Bright half/ Śuddha	2	-	-	-	-	-
12.	703	-	-	Kārtika	-	-	-	-	-	-	Śailāditya, a scion of the Guhila family of Mēwār

13	14	15	16	17
-	-	-	-	Māndasūr Insc. of the time of Naravarman; The Māḷava yr 461; EI XII, No. 35, p. 315.
-	To record the construction of a Stūpa, **Kūpa (Well)**, a prā (**Chariatable water** stall) and an **Ārāma (Garden** or Monastery)	-	467-468 A.D.	Māndasūr Insc. of Māḷava Samvat 524; EI XXVII No. 4, p. 12.
-	-	-	491 A.D.	Two Inscs. of Gauri (1) Chhōti Sadri Insc, year 547; EI XXX, No. 22, p. 122.
-	To record the rebuilt of a dilapidated Śiva temple	-	554 A.D.	Haraha Insc. of the Reign of Īśanavarman (Vikrama Sam) 611; EI XIV, No. 5, p. 110.
-	-	-	625 A.D.	Vasantgaḍh Insc. of Varmalatta; VS 682; EI IX, No. 25, p. 188.
-	-	-	642-43 A.D.	Sakrai Stone Insc. VS 699; EI XXVII, No. 7, p. 32.
-	It records that a Mahājana headed by Jentaka started a mine (Akāra) in Aryankūpagiri	Which becomes a source of lovely hord for the people	646 A.D.	Samōli Insc of the Time of Śilāditya (Vikrama Sam) 702; EI XX, No. 9, p. 97.

1	2	3	4	5	6	7	8	9	10	11	12
13.	718	-	-	Mārgaśira	Bright Half/	5	-	-	-	-	Aparājita, a king of Guhila family
14.	767	-	-	-	-	-	-	-	-	-	Pāśupata Ascetic Danarāsi during the time of Nannapa of Rāṣtrakūṭa family

13	14	15	16	17
-	Installation of the statue of Vāsudēva (Viṣṇu)	-	No.2, 661 AD	Udaipūr Insc. of Aparājita (Vikrama) Sam 718; EI IV, No. 3, p. 29
-	To record the construction a temple of Śiva	-	710-711 A.D.	Indragaḍh Insc. of Nannappa, VS 767; EI XXXII, No. 13, p. 113.

The Saka Era (78 A.D.)

Sl. No.	Samvatsara	Āyana	Season/ Ṛtu	Month/ Māsa	Fortnight/ Pakṣa	Day/ Tithi	Star/ Nakṣatra	Karaṇa	Yōga	Muhūrta	Donor
1	2	3	4	5	6	7	8	9	10	11	12
1.	60 Sabtsara	-	-	-	-	-	-	-	-	-	Rāja Namijada
2.	60 sam	-	-	-	-	-	-	-	-	-	Śivarakṣitasa
3.	72	-	-	Mārgaśira	Dark Half/ Bahuḷa	Prati, 1	-	-	-	-	The Mahākṣātrapa Rudradaman
4.	169 Prabhava	-	-	Phālguṇa	Dark Half/ Bahuḍa	Amāvāsya Bṛgu Vār	Rēvati	Vṛddhi	-	Vṛshab Lagn	-
5.	261 Viḷamba	-	-	Kārtīka	Bringt Half/Śuddha	Trayōdaśi	Aśvini	-	-	-	-

Donee	Object of the Inscription	Purpose of the Inscription	Corres Christian Era	Remarks or References
13	14	15	16	17
-	-	For the incr. of relatives and frineds (and) for the welfare of his son	-	Shahdaur Inscs. of the yr. 60; EI XIX, 30, p. 197.
-	-	-	-	*Ibid.* p. 200.
-	Restoration of the lake Sudarśana	-	Nov. 16 150 A.D.	Jūnagaḍh Rock Incs. of Rudradaman, the yr. 72; EI VIII, No. 6, p. 36.
-	-	-	Nov. 11 Friday 248 A.D.	Tānjōre Copper Plate Insc of the Western Gānga king Arivarman (Harivarman); VIII, p. 213 and Vol XVII, p. 241.
-	-	-	Oct. 13, Friday 338 A.D.	Mudyanūr Copper Plate Insc. of the Bāṇa king Śrī Vaddhu-vallabha-Malladēva-Nandivarman; IA XV, p. 175 and XVII, p.

1	2	3	4	5	6	7	8	9	10	11	12
6.	261 Vibhava	Uttara	-	Puṣya	Dark Half/Chatudasi Bahuḷa Sōmavāra		-	-	-	-	-
7.	366 Tāraṇa	-	-	Phālguna	Dark Half/Amāvāsya Kṛṣna Pak Bṛhavāra		-	-	-	-	-
8.	388	-	-	Māgha	Bright Half/ Panami Śuddha Sōmavāra		Swāti	-	-	-	-
9.	411 Vibhava S. Nṛpagbdēśa	-	-	Vaiśākha	Bright Half/ Pūrṇa- śuddha punya divase		-	-	-	L.Eclps Chandra Grahaṇa	-
10.	415 S.Nṛpakāla	-	-	Jyēṣṭha	Dark Half/Amāvāsya Kṛṭna pak		-	-	-	S.Eclps Surya Grahaṇa	-
11.	417 S. Nṛpakāla	-	-	Jyēṣṭha	Dark Half/Amāvāsya Kṛṣna pak		-	-	-	S. Eclps Sūrya Grahaṇa	-

13	14	15	16	17
				239
-	-	-	-	On the dates of the Śaka Era in inscs. IA XXIV, p. 11 (date in the Kalbhavi Jain Insc. Ante Vol XVIII, p. 311.
-	-	-	Feb. 22, Thursday, day, 445 A.D.	On the dates of the Saka Era in Inscs. IA XXIV, p. 9 (Bangaore Cop. Plate Insc. of Vira Nonamba, IA VIII, p. 95.
-	-	-	Jan. 7, Friday, 466 A.D.	IA XXIV, p. 11, SL. No. 169(Merkara Cop. Plate Insc. of Western Gānga King Avinita Kongani, IA VIII, p. 95.
-	-	-	Apr. 12, 488 AD	On the dates of the Śaka Era in insc. IA XXIV, p. 10 Sl. No 164 (Cop. Pi. Insc. of the EArly Caḷukya Pulakēṣin I; IA VII, p. 212
-	-	-	June 10, 941 A.D.	IA XXIV, p. 11, Sl. No. 170, (Bagumra Cop Pl. Insc. of Gūrjara Doḍḍa II, Praśāntarāga, IA XVII, p. 200; IA XVIII, p. 92.
-	-	-	April 21, 494 A.D.	IA XXIV, p. 10, Sl. No. 165, (Ilao Co. Pl. Insc. of Gūrjara Doḍḍa II, Praśāntaāga, IA XIII, p117; & IA XVIII, p. 92 & 286.

1	2	3	4	5	6	7	8	9	10	11	12
12.	465 S. Varṣēsu	-	-	-	-	-	-	-	-	-	-
13.	534 S. N. Samvat	-	-	Bhādrapads	Dark Half/ (of Pūrṇimānta)	15	-	-	-	S. Eclps Sūrya Grahaṇa	-
14.	552	-	-	-	-	-	-	-	-	-	The Satyāśraya (Pulakēśin II)
15.	552	-	-	Śravaṇa	-	Amāvāsya/ New Moon	-	-	-	Solar Eclps	Mahārāja Indravarman Gānga
16.	553	-	-	Mahākārtīk	-	-	-	-	-	Solar Eclps	Nanarāja, the Rāṣṭrakūṭa king
17.	556 (Kali 3735)	-	-	-	-	-	-	-	-	-	Ravikīrti found a temple of Jain Prophet Jinēdra
18.	577 Ānanda	- -	- -	Māgha	Bright Half/ Śuddha	3	-	-	-	-	King Nikumballa Śakti of Sendraka dynasty

13	14	15	16	17
-	Made the great hill of Vātāpi into invincible citadel	For the prosperity (Security) of the earth	543 A.D.	Bādāmi insc. of CÇ-ukya Vallabhēśwara; Śaka 465; Ei XXVII, No. 2, p. 5
-	-	-	August 12, 611 A.D.	IA XXIII, P 130,Si. No. 106; on the dates of the Śaka Era in Inscs. (Hyderabad Cop. Pl. Insc. of the western Cāḷlukya Pulakēśin II; IA VI, P 73; IA XVI, P 109; XVII p. 109 ; XVII 141 & XX, P 3
Damadīkṣita Sāvarṇi gōtra	Grant of the vill. Gōviymaka	-	630 A.D.	Lōhanēr plates of cāḷukya Pulakēśin II; EL XXVII, No. 9, p. 37.
To a Brāhmin	Tōṭāvaṭaka vill. was given as an Agrahāra	For the welfare of his parents and his family	630 A.D.	Andhavarman Plates of Indravarman, EI XXX No. 9, p. 38.
Muṇḍibhaṭṭa	Records a grant of land in the Tivarkhēḍa Village	-	Oct. 631 A.D.	Tivarkhēd plates of the Rāṣṭrakūṭa Nannarā-ja; Śaka 553; EI Vol XI, No. 27, p. 276.
-	-	-	634-35 A.D.	IA XXIV, p. 183, Si. No. 21; on the dates of Śaka Era in Inscs. S 556 (3735); (Aihōle Incs. of the Western Cāḷukya Pulakēśin II; IA, V, p. 70; VIII, p. 242; Arch. Sur. of W. India Vol III, p. 135; EI Vol VI, No. 1, p. 1.
Bōgika, Brāhmin	Grant of the Village Susehirakhōṭi	To gain merit for himself as well as his parents	January 15, 655 A.D.	Nagad plates of Sendruka Nikumballa Śakti, Śaka yr. 577; EI XXVII, No. 34, p. 195.

1	2	3	4	5	6	7	8	9	10	11	12
19.	589 Sakanam	-	-	Mādhava Vaiśākha	Dark Half/ Bahuḷa of Amānta	16 day (1tithi)	Anūrādha	Jupiter in Chāpa Dhanus	- (Mēṣa)	Sun in Aja	-
20.	614 S. Vrṣēṣu	Dakṣināyana	-	Puṣya	Saturday	-	Rōhiṇi	-	-	-	-
21.	596 S. Varṣēṣu Vijayarājya Saṃvatsara	-	-	Vaiśākha	Bright Half/ śuddha	Full Moon Pūrṇima	-	-	-	-	Vikramāditya I, the Western Cāḷukya king
22.	602	-	-	Phālguṇa	Dark Half/ Kriṣṇna pak	10	-	-	-	Enhtrance of Sun in Meena	Jayaśakti of Senīṛuka dynasty
23.	609	-	-	-	-	-	-	-	-	-	Vinayāditya of Early Cāḥukya dynasty
24.	613	-	-	Vaiśākha	Brigh Half/ śuddha	Full Moon Pūrṇima	-	-	-	-	Vinayāditya of Western Cāḷukya Navasari branch
25.	612 Corres. to 10th Reg. yr	-	-	Vaiśākha	Bright Half/ śuddha	Full Moon Pūrṇima	-	-	-	-	Vinayāditya of Western Cāḷukyas

13	14	15	16	17
-	-	-	April 15, 667 A.D.	IA XXIII, p. 122, Si. No. 58 (Stone Insc. of Vat pray Viar S 589; IA Ante. Vol XXI, p. 48).
-	-	-	June 22, 692 A.D.	IA XXIV, On the dates of Śaka Era in Insc. p 9, Si. No. 160 (Sorab Cop. Pi. Insc. of the W. Cāḷukya Vinayāditya IA XIX, P 150
To three Brāhmins	-	-	April 25, 674 A.D.	Gadwāl Plates of Vikaramāditya I, 674 A.D. ; EI X, No. 22, p. 101
Bappaswāmin, a Brāhmin	The gift was given	-	February 17, 681 A.D.	Muṇḍakhēḍa Plates of Senḍraka Jayaśakti; Śaka 602; EI Vol XXIX, No. 15, p. 117.
Allaśaramn,	To record the gift of the Village called Vīra	-	687A.D.	Jejuri plates of Vinayāditya; Śaka samvat 609; EI XIX, No. 8, p. 62.
-	To record the grant of some villages and domestic sites	For the benefit of the temple of Sun God at Mānapura	691-692 A.D.	Minor plates of Vinayāditya; Mangalarasa; Śaka 613; EI Vol XXVIII, No. 3, p. 17.
Durgaśarman and Raviśarman, the Brāhmins	Grant of a field in vill. Toravi	At the request of the Queen	Apr 29, Friday 69 A.D.	Poona plates of Cāḷukya Vinayāditya; Śaka 612; EI Vol XXVIII, No. 3, p. 17.

1	2	3	4	5	6	7	8	9	10	11	12
26.	614 12th Reg ye	Dakṣi-nayana	-	Aṣāḍha	Bright Half/ śudda	Full Moon Pūrṇima	-	-	-	-	Vinayāditya of Satyāsraya of W. Caḷukyas
27.	615	-	-	Kārtīka	Bright Half/ śuddha	Full Moon Pūrṇima	-	-	-	-	Nannarāja of Rṭakūṭa Dynasty
28.	621 3rd Reg yr.	-	-	Jyēṣṭha	Bright Half/ śuḍdha	Full Moon Pūrṇima	-	-	-	-	Vinayavati, mother of Vijayāditya Satyāśraya
29.	625	-	-	-	-	-	-	-	-	-	Vijayāditya, the Early Cāḷukya
30.	632 14 Reg yr.	-	-	Kārtīka	Bright Half/ śuddha	Full Moon Pūrṇima	-	-	-	-	Vijayāditya, the Early Cāḷukya
31.	653	-	-	Vaiśākha	Bright Half/ śuddha	Full Moon Pūrṇima	-	-	-	-	Vijayāditya, the Early Cāḷukya

13	14	15	16	17
To three Brāhimns	To register the grant of 50 Nivartanas of land each	At the request of a certain Mardakari	July 4, 692 A.D.	Dayyandinne plate of Vinayāditya Satyāśraya ; Śaka 614; EI Vol XXII, No. 7, p. 24.
Haragaṇadvivēdi, a Brāhmin	Gift of lands in the vills. Vimbarika and Vaṭapuraka	-	693 A.D.	Sangaluḍa plates of Rāṣṭrakūṭa Nannarāja; Śaka 615; EI Vol XXIX, No. 4, p. 110.
-	Records the installation of the Gods, Brahma, Viṣṇu, Śiva at the cap. Vaṭapyadhistāna	For the benefit of the temple of the triad	May 20, 699 A.D.	Pillar in the Jambulinga temple; Karṇāṭaka Insc. Vol I, Ed. by RS Panchamuki, Kannaḍa Research Office; 1941, p. 2.
To a Brāhmin	Record the grant of the villages Jalla and Nirlana	For the religious merit of his parents	-	Rāyagadh plates of Vijayāditya; Śaka Sam 625; EI X, No. 2, p. 14
Bhānudēva śaramn, a Brāhmin	Records the donation of a villages named Karvva	- XXVI, p. 322.	-	A Cop. Plate grant of Cāḷukya Vijayaditya; Śaka 632; Satāra Museum Plate; EI Vol
Bhavaswāmibaṭṭa	The donation of a village	-	-	A grant of the W. Cāḷukya Vijayāditya; Śaka 653; EI Vol. XXV, No. 3, p. 21.

The Cēdi/Kālacuri Era (249 A.D.)

Sl. No.	Samvatsara	Ayana	Season/ Ṛtu	Month/ Māsa	Fortnight/ Pakṣa	Day/ Tithi	Star/ Nakṣatra	Karaṇa	Yōga	Muhūrta	Donor
1	2	3	4	5	6	7	8	9	10	11	12
1.	207	-	-	Vaiśākha	Bright half/ Śuddha	Trayō-daśi	-	-	-	-	Mahārāja Dahrasēna of Trikūṭaka family
2.	241	-	-	Kārtika	Bright half/ Śuddha	Trayō-daśi	-	-	-	-	Mahārāja Vyāgrasēna
3.	292	-	-	Mahākārtik	Bright half/ Śuddha	15	-	-	-	-	Mahārāja Sangamasimha, a feudatory of Kālacuris
4.	346	-	-	-	-	-	-	-	-	-	-
5.	391	-	-	Vaiśākha	Dark half/ Bahuḷa	15/ New Moon	-	-	-		Raṇagraha, son of Vītarāga
6.	360	-	-	Bhādrapada	Bright half/ Śuddha	Trayō-daśi	-	-	-	-	Buddharāja, s/o Śanka-ragaṇa of K.C. family

Donee	Object of the Inscription	Purpose of the Inscription	Corres Christian Era	Remarks or References
13	14	15	16	17
To a Brāhmin	Grant of land in the Kanyastada-kasarika	-	Apr. 4, 456 A.D. (Current)	Pardi plates of Dahrasēna; The year 207, EI., X, No. 13, p. 51-54.
To a Brahimin, Nāgaśarman	Records a grant of land in Purōhitapālika in Ikṣaraki District	In order to increase the religious merit of his mother and father and the king himself	490-491 A.D.	Surat plates of Vyāgrasēna; The year 241, EI, XI, No. 21, p. 219-21.
To several Brāhmins	Records the gift of village Sōnavva	-	540-541 A.D.	Sunao Kala plates of Sangamasimha; (Kālacuri) Samvat 292; EI, Vol. X, No. 16, pp. 72-75.
	-	-	595-596 A.D.	A Gūrjara grant of Sam 346; EI, II, No. 14, p. 19-20.
Ādityaśarman, a Brāhmin	-	-	640 A.D.	A grant of Ranagraha son of Vītarāga; Sam 391; EI, II, No. 14, p. 20.
Bōdhaśvāmin	-	To provide for the 5 sacrifices, Bali, Cāru, Vaiśvadēva, Agnihōtra and for others	11 Aug.Friday, 607 A.D. or 29 Aug. Tuesday, 608 A.D. or 19 Aug. Tuesday, 609 A.D.	Vadnar plates of Budharāja (Kālacuri) Sam 360, EI, Vol. XII, No. 7, pp. 31-36.

1	2	3	4	5	6	7	8	9	10	11	12
7.	361	-	-	Kārtika	Dark half/ Bahuḷa	15	-	-	-	-	Buddharāja, s/o Śankaragana of K.C. family
8.	392	-	-	Vaiśākha	Bright half/ Śukla	15/ Full Moon	-	-	-	-	Daḍḍa IV, Praśāntarāga
9.	404	-	-	Āṣāḍha	Dark half/ Bahuḷa	15/30/ Amāvāsya	-	-	-	S.Eclips/ Sūrya-Grahaṇa Kāla	Nikumbala Śakti of Sendruka family
10.	421	-	-	Māgha	Bright half/ Śuddha	13	-	-	-	-	Sryasrayaśilāditya s/o Dharasraya, the Western Cāḷukyan king of Bādāmī
11.	-	-	-	Aśvina	Dark Half/ Bahuḷa	11	-	-	-	Tulā Sankrānti	Gūrjara Jayabhaṭṭa III

13	14	15	16	17
Bappaswamin, a Brāhmin	Granted the village Kumarivāḍa	-	609-610 A.D.	Sarsavni plates of Buddharāja (Kalācuri) Sam 361, EI, Vol. VI, No. 29, pp. 294-295.
Sūrya, a Brāhmin	To convey two fields-one at Suvarṇarapalli and one at Kshirasāra in the Sangama Khēṭaka Viṣaya	For defraying the expense of his sacrifices	641-642 A.D.	Two grants of Daḍḍa IV, Praśāntarāga (Cedi Sam 392; EI, Vol. V, No. 5, pp. 37-41.
Balapravasita	50 Nivartanas of land at the vill. Pippalakhēṭa for the upabhōga of God Langhyēśwara	To acquire merit for himself and his parents	June 1, 653 A.D.	Two Sendruka plates (B) Kasara plates of Nikumbala Śakti. Year 404; EI, Vol. XXVIII, No. 34, p. 196.
Advaryu, the pupil of Kikkaswamin	The donation of Asattigrāma	-	671-672 A.D.	Navsāri plates of Sryasrayasiladitya, the year 421; EI, Vol., VIII, No. 22, pp. 230-233.
Nārayaṇa, a Brāhmin	60 Nivartanas of land in and arround the village Tauranaka in Nandipura Vishaya.	For the merit of his parents and for himself	709-710 A.D.	Anjanērī Plates of Gūrjara Jayabhaṭṭa III, K 461; EI, Vol. XXV, No. 31, p. 292-294.

The Gupta Era
(319 A.D.)

Sl.. No	Samvatsara	Ayana	Season/ Ṛtu	Month/ Māsa	Fortnight/ Pakṣa	Day/ Tithi	Star/ Nakṣatra	Karaṇa	Yga	Muhūrta	Donor
1	2	3	4	5	6	7	8	9	10	[illegible]	12
1.	9	-	-	Vaiśākha (April-May)		10 Solar	-	-	-	-	Samudragupta
2.	61	-	-	Āṣāḍha (June-July)	Bright Half/Śukla	5	-	-	-	-	Candragupta II, the s/o Samudragupta
3.	66	-	-	Mārgaśira (Nov-Dec)	Bright Half/Śukla	7	-	-	-	-	Adityasena
4.	82	-	-	Āṣāḍha (Jun-Jul)	Bright Half/Śukla	11	-	-	-	-	A Mahārāja of Samkanika tribe, a feudatory of Candragupta II

Donee	Object of the Inscription	Purpose of the Inscription	Corres. Christian Era A.D.	Remarks of References
13	14	15	16	17
To a Brāhmin	The vill. of Revatika in the Gaya Vishaya was granted	-	328-329	Gaya Cop.Pl.Insc. of Samudragupta the year 9; Corpus of the Inscriptionum Indicarum, Vol.III, Gupta Inscs.No. 60, Plate XXXVII p. 255.
-	The statues of two teachers were installed in the teachers shrine	For baseeching the worshipper of Mahesvara	380	Madhura Pillar Insc. of Candra Gupta II, GE 61; EI XXI, No. 1, p. 1.
-	-	The Insc. is one of the Solar worship	672-673	Shahpur stone image Insc. of Adityasena; No. 43, plate XXIX, A, p. 209 of Corpus of the Inscriptionum Indicarum Vol.III Gupta Inscriptions
-	To record the gift	Vaisnava Inscription	401-402	Udaigiri Cave Insc. of Candra Gupta II, the year 82, No. 3. Pl.II B, Corpus of the Inscription Indicarum, Vol.III, Gupta Inscriptions, p. 23

1	2	3	4	5	6	7	8	9	10	11	12
5.	88 Reg Yr. of C. Gupta	-	-	-	-	-	-	-	-	-	Candragupta II
6.	93	-	-	Bhādrapada (Aug-Sept)	Without any Specn. of L. Fortnight	4	-	-	-	-	Amrakardhava, an officer of Candragupta II
7.	96 Regnal year of Kumara Gupta	-	-	-	-	-	-	-	-	-	Dhruvasarman
8.	98 Regnal year of Kumara Gupta	-	-	-	-	-	-	-	-	-	-
9.	106	-	-	Kārtika (Oct-Nov)	Dark Half/ Kṛṣṇa Pakṣa	5	-	-	-	-	-

13	14	15	16	17
To a satra/charit-able hall	Records the gift of 10 dinars	To the perpetual manienance of a satra	407-408	Gadwa Stone Insc. of Candra Gupta II, the year 88, No. 7, Pl.IV B, Corpus of the Inscriptionum Indicarum, Vol. III, Gupta Inscriptions, p. 137.
To the Aryasangha	To record the grant of a vill. Isvaravasaka and a sum of money	For feeding mendicants and Maintaining lamps	412-413	Sanchi Stone Insc. of Candra Gupta II, the year 93, No.5, Pl.III B, Corpus of the Inscriptionum Indicarum, Vol. III, Gupta. Inscriptions, p. 29
	Accomplishment of works viz., Bldg. of a gateway with a flight of steps, estb. of satra, erection of a column with insc. at Kartikeya temple	-	415-416	Bilsad Stone Insc. of Kumara Gupta, year 96, No. 10, Pl.V, Corpus of the Inscriptionum Indicarum, Vol.III, Gupta Inscriptions, p. 43
	Gift of 12 dinars as contbn. to the perpetual maintenance of a satra/charitable hall	-	417-418	Gadhwa Stone Insc.of Kumara Gupta, the year 98, No. 9, Pl.IV, D, Corpus of the Inscriptionum Indicarum, Vol. III, Gupta Inscriptions, p. 40.
	The installation of an image of the Tirthankara Parsva at a cave	-	425-426	Udaigiri Cave Insc., the year106, No. 61, Pl. XXXVIII, A, p. 258 Corpus of the Inscriptionum Indcarum, Vol. III, Gupta Inscriptions

1	2	3	4	5	6	7	8	9	10	11	12
10.	113	-	-	-	-	-	-	-	-	-	Kumara Gupta
11.	116	-	-	-	-	-	-	-	-	-	By 5 brothers
12.	117	-	-	Kārtika (Oct-Nov)	-	10	-	-	-	-	Pṛthvisēna, a minister of Kumāra Gupta
13.	121	-	-	Vaiśākha (Apr-May)	-	1	-	-	-	-	A board of admn.
14.	128	-	-	Māgha (Jan-Feb)	-	19	-	-	-	-	By two persons, Bhōyila and Bhāskara
15.	129	-	-	Jyēṣṭha (May-Jun)	-	18	-	-	-	-	Kumāra Gupta
16.	131	-	-	Āśvayuja (Sep-Oct)	-	5 Solar	-	-	-	-	Hariswāmin of the time of Kumāra Gupta/Skanda Gupta

13	14	15	16	17
Varāhaswāmin, a Sāmavēdi Brāhmin	-	-	432-433	Dannaidaha Cop. Pl. Insc. of the time of Kumāra Gupta, the yr. 113, EI Vol. XVII, No. 23, p. 345.
-	To record the construction of a temple of a God whose name is lost	-	435	Tumain Insc. of Kumāra Gupta and Ghaṭōtkacha Gupta; GE 116; EI. Vol. XXVI, No. 11
-	Gift was made for the worship of Mahādēva, known as Pṛthvīśvara	-	436	Karamdanda Insc. of the reign of Kumāra Gupta (Gupta), Sam 117; EI. X, No. 15, p. 70.
To the Brāhmaṇas	Records the donation of uncultivated land	-	440	Sultānpūr Cop. Pl. Insc.; EI. XXXI, Na 9, p. 57.
To the Gōvinda-swāmy temple	Purchase of 3 Kulavapas of revenue free khila fields and two dīnārs of stalavastu state	To meet the expenses of flowers & perfumes etc. reqd. for daily worship of the God	447-448	Baigrām Cop. P1.Insc. of the (Gupta) year 128; EI. Vol. XXI, No. 13, p. 78.
-	The instalation of the image on the pedastal of Insc.	Sarvadukhapariharaṇārdham	448-449	Mankuwar Stone Image Insc. of Kumāra Gupta, yr. 129, No. 11, Pl. VI, Corpus of the Inscriptionum Indicarum .Vol. III, Gupta Insc. p. 46.
To the Āryasangha at the great Vihāra or Buddhist convent kakanaḍabṛta	The grant of certain sums of money	Feeding one Bikshu day by day and of maintaining lamps in the shrines of Buddha	450-451	Sānchi Stone Insc., the yr 131, No.62, Pl. XXXVIII, B, p.261 Corpus of the Inscriptionum Indicarum, Vol. III, Gupta Insc.

1	2	3	4	5	6	7	8	9	10	11	12
17.	135	-	-	Pu,ya (Dec-Jan)	-	20 Solar	-	-	-	-	-
18.	136	-	-	Pauṣṭapad (Aug-Sep)	-	6	-	-	-	-	-
	137	-	-	-	-	-	-	-	-	-	-
	138	-	-	-	-	-	-	-	-	-	Chakrapālita
19.	139	-	of a Season	2nd month	2nd fortnight	7 Solar	-	-	-	-	Bhīmavarman, a feudatory of the Early Gupta king Skandagupta

13	14	15	16	17
-	Gift of the statue on the pedastal of which the insc. is engraved	-	454-455	Madhura Stone Image Insc. the yr.135, No. 63, Pl.XXXIX, A, Corpus of the Inscriptionum Indicarum Vol.III, Gupta Insc.p. 263
-	To record that making the Calcn. in the reckoning of the Guptas in the yr 137 at	-	455-456	
-	night lake Sudarśana burst in due to excessive rain.	-	456-457	Junagadh Rock Insc. of Skandagu the years 136, 137, 138, No. 14, Plate, VIII, p. 56. Corpus of the Inscriptionum Indicarum Vol. III, Gupta Insc.
	The restoration of breach after two months work in 137.			
-	It goes on record in the 138th year Chakrapālita built a temple of Vishnu under the name of Chakrabṛt.	-	457-458	-
-	The gift or installation of the sculpture on the base of the inscription.	-	458-459	Kosam Stone Image Insc. of M.Bhīmavarman; the year 139; No. 65, plate XXXIX C; p. 226; Corpus of the Inscriptionum Indicarum Vol.III, Gupta Insc.

1	2	3	4	5	6	7	8	9	10	11	12
20.	141	-	-	Jyēṣṭha (May-Jun)	-	-	-	-	-	-	A certain Madra
21.	146	-	-	Phālguna (Feb-Mar)	-	-	-	-	-	-	Dēvaviṣṇu, a Brāhmin
22.	148	-	-	Māgha (Jan-Feb)	-	21 Solar	-	-	-	-	-
23.	156	-	-	Kārtika (Oct-Nov)	Bright half/ Suddha	3 Lunar	-	-	-	-	Mahārāja Hastin
24.	158	-	-	Jyēṣṭha (May-Jun)	Bright half/ Suddha	Full Moon Pūrṇima	-	-	-	-	Mahārāja Lakṣmaṇa

13	14	15	16	17
-	To record the set up of 5 stone images of Adhikartris	-	460-461	Kahaum Stone Pillar Insc. of Skandagupta, the yr. 141, No. 15, Plate IX A, p. 66. Corpus of the Inscriptionum Indicarum. Vol.III Gupta Insc.
-	A perpetual endowment in a temple of Sun at Indrapura	For maintaining a lamp in the Temple	465-466	Indore Cop.Pl.Insc. of Skandagupta the yr. 146, No. 16, Pl. IX B, Corpus of the Inscriptionum Indicarum. Vol. III, Gupta Insc. p. 68.
-	The installation of an image of the God Viṣṇu under the tree of Anantaswamy	Providing perfumes, incense, Garlands etc. and of executing repairs	467-468	Gadhwa stone Insc. the yr. 148, No. 66, Pl. XXXIX D, P. 267 Corpus of the Inscriptionum Indicarum. Vol.III Gupta Insc.
Gōpasvāmin and other Brāhmins	Grant of the vill.of Vasundhara Shanḍika	-	475-476	Corpus of the Inscriptionum Indicarum vol.III, Gupta Insc. Khoh Cop.Pl.Insc. of the Mahārāja Hastin; the year 156; No.21, No. Plate; p. 95.
Raviśarman a Brāhmin	Grant of an Agrahāra in the vill. Phēlaparvatīya, (Pāli)	-	-	Pāli Cop.Pl. grant of Mahārāja Lakṣmaṇa; Samvat 158; EI. Vol.II, No. 30, p. 363.

1	2	3	4	5	6	7	8	9	10	11	12
25.	159	-	-	Māgha (Jan-Feb)	-	7 day	-	-	-	-	A private individual
26.	163 Mahāśvayuja Samvatsare	-	-	Chaitra (Mar-Apr)	Bright half/ Suddha	2 Lunar	-	-	-	-	Mahārāja Hastin
27.	165	-	-	Aṣāḍha (Jun-Jul)	Bright half/ Suddha	12 Lunar Sūrya-guru Divase	-	-	-	-	Mahārāja Matriviṣnu and his younger brother, Dhānyaviṣnu
28.	167	-	-	Bhādrapada (Aug-Sept)	Bright half/ Suddha	7 Lunar	-	-	-	-	Mahārāja Subandhu
29.	174	-	-	Aṣāḍha (Jun-Jul)	Not specified	14 Lunar	-	-	-	-	Mahārāja Jayanātha

13	14	15	16	17
To a Brāhmaṇa couple for the worship of Jinas	Registers the purchase of a fallow land	For charitable purposes	Jan, 479	Paharpūr Cop. Pl. Grant of the (Gupta) year 159; EI. Vol. XX, No. 5, p. 59.
Certain Brāhmaṇas	The grant of the Agrahāra of Kōrparika	-	482-483	Corpus of the Inscriptionum Indicarum. Vol.III, Gupta Insc. Khoh Cop. Pl.Insc. of the Mahārāja Hastin; the year 163; No. 22, Plate XIII, p. 100.
-	The erection of the column called Dvajastambha/Flag staff of God Viṣṇu under the name of Janārdhana	-	484-485	Corpus of the Inscriptionum Indicarum Vol.III, Gupta Insc. Eran stone pillar Insc. of Budhagupta; the year 165; No.19, Plate XII A, p. 89.
Ṣaṣṭisvāmin, a Brāhmin	The grant of a piece of land at the vill.Sohajana in the Udumbhagarta	-	486	The Barwanai Cop.Pl.Insc.of Mahārāja Subandhu; the year 167, EI.Vol.XIX, No. 44, p. 261.
To a Brāhmin	Grant of the vill. Chchandapallika	-	493-494	Corpus of the Inscriptionum Indicarum. Vol.III, Gupta Insc. Karitalai Cop. Pl. Insc. of Mahārāja Jayanātha; the year 174; No. 26, Plate XVI, p. 117.

1	2	3	4	5	6	7	8	9	10	11	12
30.	177	-	-	Chaitra (Mar-Apr)	-	22	-	-	-	-	Mahārāja Jayanātha
31.	189 Mahāmāgha Sambatsare	-	-	Kārtika (Oct-Nov)	-	19	-	-	-	-	Two Mahārājas, Hastin of the Parivrajaka family & Sarvanātha of Ucchakalpa
32.	191.	-	-	Śrāvaṇa (Jul-Aug)	Dark Half/ Bahuḷa Pakṣa	7 Lunar & Solar Day of the month	-	-	-	-	Gōparāja
33.	191 Mahāchaitra Samvatsara	-	-	Māgha (Jan-Feb)	Dark Half/ Bahuḷa Pakṣa	3 Lunar & 3 Solar Day of the whole Month	-	-	-	-	Mahārāja Hastin

13	14	15	16	17
To some Brāhmaṇas	To record the grant	For the purpose of a Temple of the God Viṣṇu	496-497	Corpus of the Inscriptionum Indicarum. Vol.III, Gupta Insc. Khoh Cop. Pl.Insc. of the Mahārāja Jayanātha; the year 177; No. 27, Plate XVII, p. 121.
-	To record the erection at Ambōla of a Boundary pillar	Treaty between the two Mahārājas in question	508-509	Corpus of the inscriptionum Indicarum. Vol.III Gupta Insc Bhūmāra stone pillar Insc. of Mahārāja Hastin and Sarvnātha; No.24, Plate XV A, p. 110.
-	To record that in the Co. of king-Bhānugupta, a Chieftain named Gōparāja came to the place where the pillar was set and fought a battle	In memory of Gōparāja	510-511	Corpus of the Inscriptionum Indicarum. Vol.III, Gupta Insc. Eran Posthumous Stone Pillar Inc. of Gōparāja; the yr.191; No.20, Plate XII B, p. 92.
To certain Brāhmins	The grant of a vill. of Valugarta	At the request of a person, Mahādevīdēva	510-511	Corpus of the Inscriptionum Indicarum. Vol.III, Gupta Insc. Majhgawam Cop. Pl.Insc. of Mahārāja Hastin; the yr.191, No. 23, Pl. XIV, p. 106.

1	2	3	4	5	6	7	8	9	10	11	12
34.	191	-	-	Dvirāṣādh & (Jun-Jul)	-	10th day of inter Calary Month	-	-	-	-	Mahārāja Sarvanatha
35.	193	-	-	Caitra (Mar-Apr)	Not specified	10th day of the month	-	-	-	-	Mahārāja Sarvanātha
36.	197	-	-	Āśvayuja (Sep-Oct)	Not specified	20th day of the month	-	-	-	-	-
37.	198 Mahāsvayuja Samvatsare	-	-	-	-	-	-	-	-	-	Mahārāja Hastin
38.	207	-	-	Vaiśākha (Apr-May)	Dark Half/ Bahuḷa Pakṣa	15	-	-	-	-	Dṛvasēna
39.	214	-	-	Puṣya (Dec-Jan)	Not specified	6	-	-	-	-	Mahārāja Sarvanātha

13	14	15	16	17
To two Individuals named Viśākha-datta and Saki	Grant of the vill. of Vaiśyavātaka as an Agrahāra	For the maintenance of the temple of Kartikeya	510-511	The Sohawal Cop. Pl. Insc. of the Mahārāja Sarvanātha; yr.191; EI. Vol. XIX, No. 21, p. 127.
-	To record the grant of a vill Aśramaka on the Agreement between him and the grantees that it was applied to amongst other things	The purposes of a shrine of Vishnu under the title Bhagavat	512-513	Corpus of the Inscriptionum Indicarum. Vol.III, Gupta Insc. Khōh Cop.Pl.Insc. of the Mahārāja Sarvanātha; the year 193; No. 28, Plate XVIII, p. 126.
-	-	-	516-517	Corpus of the Inscriptionum Indicarum Vol.III, Gupta Insc. Khāh Cop. Pl.Insc. of the Mahārāja Sarvanātha; the year 197; No.30, Plate XIX, B, p. 133.
Several Brāhmins of Parāśara gōtra	Grant of the vill. called Navagrama Rashtra of the Pulinda chief	-	April 23, 517	Navagrāma grant of Mahārāja Has (G.E.(1) 98); EI.Vol.XXI, No. 20, p. 125.
A Brāhmin	To record an order of eight measures of land & two Cisterns in the vill. of Hariyanaka	-	526-527	Ganeshgadh Pl. of Dṛvasēna I (Gupta) Sam 207; EI.Vol.III, No. 46, p. 318.
	To record the sanction to the transfer among Pvt. grantees of two vills. Named Vyagrapallika & Kacharapallika	For the purpose of a temple of the Goddess Pishtapurika Dēvi at the town of Mānapura	533-534	Corpus of the Inscriptionum Indicarum. Vol.III, Gupta Insc. Khōh Cop. Pl.Insc. of the Mahārāja Sarvanātha; the year 214; No.31, Plate XX, p. 135.

1	2	3	4	5	6	7	8	9	10	11	12
40.	230	-	-	-	-	-	-	-	-	-	A Śākhya female mendicant named Jayabhatta
41.	232	-	-	Māgha (Jan-Feb)	-	20	-	-	-	-	Mahārāja Nandana, a feudatory chief of the later Guptas
42.	250	Uttara	-	Māgha (Jan-Feb)	Dark Half/ Kṛṣṇa Pakṣa	11	-	-	-	-	Mahārāja Dharmarāja, a Feudatory of Pṛthvīgraha Bhaṭṭāraka
43.	252	-	-	Vaiśākha (Apr-May)	Dark Half/ Bahuḷa Pakṣa	15	-	-	-	-	Maitraka king Dhārasēna II of Vallabhi
44.	252	-	-	Vaiśākha (Apr-May)	Dark Half/ Bahuḷa Pakṣa	15	-	-	-	-	Maitraka king Dhārasēna II of Vallabhi
45.	255	-	-	Āśvayuja (Sep-Oct)	Bright half/ Suddha	13	-	-	-	-	Sāmanta Mahārāja Simhāditya of the Garulaka family

13	14	15	16	17
To the monastery called Yashō-vihāra	To record the gift	-	549-550	Corpus of the Inscriptionum Indicarum. Vol.III, Gupta Insc. Mathura Stone Insc. the year 230; No. 70, Plate XLD, p. 273.
Raviśarman, a Brāhmin	The grant of village Mallayaṣṭika	-	551-552	A Mouna Pl. of the Mahārāja Nan (Gupta) Sam 232; EI.Vol. X, No. 12, p. 49.
-	Grant of the village called Arthakāmānduka together with another locality called Chandravātika	-	December 20,569	Sumandala Pl. of the time of Pṛthvīgraha Bhaṭṭāraka; GE 250; EI. Vol. XXVIII, No. 16, p. 79.
-	Grant of a land	-	571	Paliṭāṇa plates of Dhārasēna II; (Gupta) Sam 252; EI.Vol.XI, No.5, p. 80.
To a Brāhmin	The grant of some lands at the Vills. Antarātra, Dombi grāma, Vajragrāma	For maintenance of 5 great sacrificial rites	571-572	Corpus of the Inscriptionum Indicarum. Vol.III, Gupta Insc. Maliya Cop. Pl. Insc. of the Mahārāja Dhārasēna II; the yr. 252; No. 38, Plate XXIV, p. 165.
A Brāhmin lived at Elapadra	The grant of a field with a pond in the vill Dharbachara	-	574	Paliṭāṇa Plates of Simhāditya; yr.255; EI.XI, No. 2, p. 17.

1	2	3	4	5	6	7	8	9	10	11	12
46.	269	-	-	Chaitra (Mar-Apr)	Bright half/ śuddha	7 Solar	-	-	-	-	A certain Mahānaman
47.	286	-	-	Aṣāḍha (Jun-Jul)	Dark Half/ Bahuḷa Pakṣa	8	-	-	-	-	Parama mahēśvara śilāditya I alias Dharmāditya of Bhaṭṭāraka family
48.	289	-	-	Śrāvaṇa (Jul-Aug)	Dark Half/ Bahuḷa Pakṣa	13	-	-	-	-	14 Donors donated on their fathers' names
49.	300	-	-	-	-	-	-	-	-	Solar Eclipse	Mādhavarāja II, a Mahā Sāmanta Rāja of Śaśāṅka
50.	320	-	-	Bhādrapada (Aug-Sep)	Dark Half/ Bahuḷa Pakṣa	5	-	-	-	-	Maitraka king, Dhṛvasēna II
51.	321	-	-	Chaitra (Mar-Apr)	Dark Half/ Bahuḷa Pakṣa	3	-	-	-	-	Maitraka King, Dhṛvasēna II

13	14	15	16	17
-	To record the erection of a mansion of Buddha i.e., a Buddhist temple at the Bōdhimaṇḍala	-	588-589	Corpus of the Inscriptionum Indicarum, Vol.III, Gupta Insc Bōdhgaya Insc. of Mahāhaman; yr. 269; No. 71, Pl. XLI A;
-	-	-	605	Navalakhi Pl. of Śilāditya I (Gu Sam 286; EI.Vol.XI, No. 17, p. 175.
-	It refers to the reign of Dhruhlana, the ruling prince the donation of amount	-	608	Dadhimati-māta Insc. of the Time Dhruhlana; (Gupta) Sam 289; EI.Vol.XI, No. 31, p. 299.
Chrampasvāmin, a Brāhmin	Grant of the vill. Chavalakkhēya in Kṛṣnagiri Viṣaya	-	619	Plates of the time of Śaśāṅka Rāja (Gupta) Sam 300; EI.Vol. VI, No. 14, p. 143.
To two Brāhmins	A grant of land	-	639-640	Two grants of Dhṛvasēna II (A) Nogava Plates of (Gupta) Sam 320; EI.VIII, No. 20 A, p.188.
To two Brāhmins	The grant of a field measuring 100 Bhaktis	-	640-641	Two grants of Dhṛvasēna II (B) Nogva Plates of (Gupta) Sam 321; EI.VIII, No. 20 B, p.194

1	2	3	4	5	6	7	8	9	10	11	12
52.	334	-	-	Māgha (Jan-Feb)	Bright half/ śuddha	9	-	-	-	-	Maitraka King, Dhṛvasēna III
53.	347	-	-	Vaiśākha (Apr-May)	Dark Half/ Bahuḷa Pakṣa	15	-	-	-	-	Śilāditya III
54.	350	-	-	Phālguṇa (Feb-Mar)	Dark Half/ Bahuḷa Pakṣa	3	-	-	-	-	Śiladitya II, Vallabhi king
55.	428 Māḷava/ Vikrama	-	-	Phālguṇa (Feb-Mar)	Dark Half/ Bahuḷa Pakṣa	15 Lunar	-	-	-	-	Viṣṇuvardhana of the varika tribe, the feudatory of the Early Gupta, Samudragupta
56.	447 Māḷava/ Vikrama	-	-	Jyēṣṭha (May-Jun)	Bright Half/ Suddha	6 Lunar	-	-	-	-	Śilāditya VII (Dhrubhata) of the family of Vallabhi kings

13	14	15	16	17
Bhaṭṭibata, a Brāhmin	Grant of the vill. of Pattapadraka	-	653-654	A Vallabhi grant of Dhṛvasēna, III Sam 334; EI.,Vol. I, No. 13, p. 85.
To a Brāhmin with two names viz. Saggala and Prakāśa	Grant of 100 padavartas of land in the vill.Kukkapadra	For the religious merit of his parents	666-667	The Jesar Plates of Śilāditya 3 the year 347; EI. Vol. XI, No. 36 209.
To two Brāhmin brothers, Bhatti and Īśwara	Grant of land	For the Increase of the spiritual merit of the King's parents.	669-670	Lunsadi plates of Śilāditya I (Gupta) Sam 350; EI., Vol. IV. No.8, p. 74.
-	To record the erection of the pillar called Yūpa/sacrificial post	-	371-372 Exp. 372-373 AD.C	Corpus of the Inscriptionum Indicarum, Vol.III, Gupta Insc. Bijaygaḍh Stone pillar Insc. Viṣṇuvardhena; the yr. 428; No. 59, Plate XXXVI, C, p. 253.
To a Brāhmin	The grant of the vill. of pathaka in the Khētakahara	For the maintenance of the 5 great sacrifices and other rites	766-767	Corpus of the Inscriptionum Indicarum, Vol.III, Gupta Insc. Alina Cop. Pl.Insc. of the Śilāditya VII, the year 447; No. 39, Plate XXV, p. 172.

The Vallabhi Era (318-319 A.D.)

Sl. No.	Samvatsara	Āyana	Season/ Ṛtu	Month/ Māsa	Fortnight/ Pakṣa	Day/ Tithi	Star/ Nakṣatra	Karaṇa	Yōga	Muhūrta	Donor
1	2	3	4	5	6	7	8	9	10	11	12
1.	206 Sam	-	-	Bhādrapada	Bright half/ Śuddha	5	-	-	-	-	Mahārāja Dhṛvasēna I, a Māha Sāmanta
2.	210 Sam	-	-	Śrāvaṇa	Bright half/ Śuddha	15	-	-	-	-	Mahārāja Dhṛvasēna I, a Maha Sāmanta
3.	210 Sam	-	-	Āśvayuja	Dark half/ Bahuḷa	5	-	-	-	-	Mahārāja Dhṛvasēna I, a Mahā Sāmanta
4.	254 Sam	-	-	Vaiśākha	Dark half/ Bahuḷa	15	-	-	-	Sūryō-Parāga/ Solar Eclipse	Dhṛvasēna II of Maitraka family of Vallabhi

Donee	Object of the Inscription	Purpose of the Inscription	Corres Christian Era	Remarks or References
13	14	15	16	17
Kumāraśarman and Jharobhajin	Several plots of land in the villages Madakana, Tapasiya and Tinishaka in the Hastavapra Haraṇi	For the merit of his parents and for himself	525-526 A.D.	Paliṭāṇa Plates of Dhṛvasēna I; (Vallabhi) Samvat 206, Five Vallabhi Plates, EI, XI, No. 9; p. 105-112.
Viṣnuśarman, a Brāhmin	In order to confirm the same land in the Hastavapra Āhāra viz., Karaḍa field	In order to increase the religious merit of his mother and father and in order to be cultivated or assigns it.	529 A.D.	Paliṭāṇa Plates of Dhṛvasēna I; (Vallabhi) Samvat 206 Five Vallabhi plates, EI, XI, No. 9; p. 109-111.
Skanda, a brahmin of Aupasvasti gōtra	30 padavartas of land with an irrigation well (vāpa) was granted	To increase the religious merit of his mother and father and obtain for himself in this world and other such rewards as he may wish	529 A.D.	Paliṭāṇa Plates of Dhṛvasēna I; (Vallabhi) Samvat 206, Five Vallabhi plates, EI, XI, No. 9; p. 112.
Dēvadatta, a Brāhmin	The property consisted of a Village named Bhaṭṭaka for the maintenance of the five sacrificial rites	-	March 19, 573 A.D.	Two unpublished Vallabhi grants Bhantia Plates of Dhṛvasēna II (Vallabhi) Samvat 254, EI, XXI, No. 30 A, p. 179-181.

1	2	3	4	5	6	7	8	9	10	11	12
5.	304 Sam	-	-	Māgha	Bright Half/ Śuddha	7	-	-	-	-	Dharasēna III son of Kharagraha who was the younger Brother of Śilāditya
6.	357 Sam	-	-	2 month of Pauṣya	Dark half/ Bahuḷa	4	-	-	-	-	Śilāditya III
7.	470 Sam	-	-	Mārgaśira	Bright half/ Śuddha	7th tithi Tues-day Bhau-mavār	-	Viṣṭi	-	-	Candrāditya, a feuda-tory ruler under Kakka/Nanna at Khēṭaka

13	14	15	16	17
Mitrayāsas, a Brāhmin	100 padavartas of land in the vill. Amarakarakūpa in the Hastavapra Āhāra and a ploughed field in Dhabaka vill. In the Kalapakapataka	For the religious merit of the donor's parents	622-623 A.D.	Bhāvnagar Plates of Dharasēna III; (Vallabhi) Samvat 304; EI, XXI, No. 30 B, p. 181-184.
To a Brāhmin, who was a Dikṣita	To record the grant of a reservoir of water extending over 25 padavartas of land from the royal domain land in the Nadasara stali in the Village Madasara	For the religious merit of the donor's parents and himself	675-676 A.D.	Jesar Plates of Śilāditya III; Vallabhi Samvat 357; EI, XXII, No. 19; p. 114-120.
Bhaṭṭtamātrugana, a Brāhmin	Donation of a peace of land near the vill. of Hil□hila	For the religious merit of the donor's parents	788 A.D.	Hilōl plates of year 470; EI, XXXIV, No. 33, p. 214.

The Harṣa Era (606-607 A.D)

Sl No.	Samvatsara	Āyana	Season/ ṚTU	Month/ Māsa	Fortnight/ Pakṣa	Day/ Tithi	Star/ Nakṣatra	Karaṇa	Yōga	Muhūrta	Donor
1	2	3	4	5	6	7	8	9	10	1	12
1.	Samvat 22	-	-	Kārtika	Dark fort-night/Badi	1 Tithi	-	-	-	-	Harṣa of Kannuj and Thaneswar
2.	Savat 25	-	-	Mārgaśira	Dark Fort-Night/ Vāḍi	6 Lunar day	-	-	-	-	Harṣavardhana
3.	73-from the foundation of the kingdom Āśvayuja Samvatsara, a year of the 12 year Cycle of Jupiter	-	-	-	-	-	-	-	-	-	Mahārāja Bheṭṭi of Kiṣkinda

Donee	Object of the Inscription	Purpose of the Inscription	Corres Christian Era	Remarks or References
13	14	15	16	17
Bālacandra & Badraswāmin, the Brāhmins	The village was granted in Markaṭasāgara	For the religious merit of his parents	628-629 AD	Banskhera Plate of Harṣa; EI IV, No.29, p. 208.
Two Brāhmins	The village Sōmakuṇḍaka in the Kundabani Viṣaya which had been previously held by a Brāhmin on the strength of a forged charter was granted to	For the increase of the merit and fame of his ancestors and his name	631-632 A.D. Nov-Dec	The Madhuban Copper plate of Harṣa dated Samvat 25; EI, I, No. 11, p. 67-75 and EI, VII, No. 22, p. 155-158.
To the Brāhman Baṭṭināga	To record the gift of the Agrahāra village Ubbāraka	For the religious merit of Mahārāja Bhetti who may have been the King's father	679-680 A.D.	Dhulev Plates of Mahārāja Bheṭṭi, year 73, EI. XXX, No. 1, p. 1-2.

1	2	3	4	5	6	7	8	9	10	11	12
4.	88-26th Regnal year	-	-	Kārtika	Bahuḷa/ Dark	-	-	-	-	-	Madhyama Rājyadēva of Śailōdbhava family
5.	100	-	-	Phālguṇa	Bright half/Su	13 day	-	-	-	-	Bhōjadēva, a Mahārāja
6.	100	-	-	Vaiśākha	Bright Half/Su	5 day	-	-	-	-	Subhakara II, (Bhaumakara) king of Orissa

13	14	15	16	17
To 12 Brāhmins	To record the grant of a vill. in the Katakabhukti Visaya	For the religious merit of his parents	694 A.D.	The Parikud Plates of Madhyama Rājyadeva; EI.XI, No. 28, p. 281-286.
-	The grant of the vill. of Śiva in the Denḍavanaka Viṣaya	Renewal of the grant which had been by his great grandfather, Mahārāja Vatsarājadēva	706 A.D.	Daulatpura Plate of Bhōjadēva I of Mahōdaya (Harsha) Samvat 100; EI.V, No. 24, p. 208-21
6 Brāhmins of Bharadvāja Gōtra	As a free revenue permanent gift of a vill. Lavagunḍa on request of the queen Nṛnna	-	-	Tērunḍia Plate of Subhakar all (Bhaumakara); EI, Vol. XXVIII, No. 36, p. 211-216.

Regnal Years : The Mauryas (320 B.C.-185 B.C.)

Sl. No.	Samvatsara	Āyana	Season/ Ṛtu	Month Māsa	Fortnight/ Pakṣa	Day/ Tithi	Star/ Nakṣatra	Karaṇa	Yōga	Muhūrta	Donor
1	2	3	4	5	6	7	8	9	10	11	12
1.	40th year	-	Hēmanta	-	1st Pakhe	10th divase	-	-	-	-	Aśōka, the Maurya
2.	21st year of his reign	-	-	-	-	-	-	-	-	-	Aśōka
3.	256th year	-	-	-	-	-	-	-	-	-	Aśōka
4.	13th year	-	-	-	-	-	-	-	-	-	Aśōka

Donee	Object of the Inscription	Purpose of the Inscription	Corres. Christian Era A.D.	Remarks or References
13	14	15	16	17
-	-	-	249 B.C.	Epigraphical Discoveries at Sarnath; EI., Vol. VIII, p. 166-171.
-	The king remitted the taxes of the village Lummini	-	259 B.C.	The Aśōka Edicts of Pedaria and Nigliva; EI, Vol. V, No. 1, p. 1-4.
-	-	-	-	Ahraura Insc. of Aśōka, EI., XXXVI, No. 30, p. 139.
-	-	-	256 B.C.	Aśōkan Edicts by Amūlya Candrasēn, The Indian Pub., Society, 1956, p. 78.

Regnal Years : The Śātavāhanas (75 B.C.-250 A.D.)

Sl. No.	Samvatsara	Āyana	Season/ Ṛtu	Month/ Māsa	Fortnight/ Pakṣa	Day/ Tithi	Star/ Nakṣatra	Karaṇa	Yōga	Muhūrta	Donor
1	2	3	4	5	6	7	8	9	10	11	12
1.	6 sava	-	Summer/ Gīmha	Vaiśākha	4 Pakhe	Pūrṇima/ Punnima	-	-	-	-	Gautamīputra Vijaya Śātakarṇi
2.	18 Savachhara	-	Rainy/ Vasa	-	2 Pakhe	1 Divase	-	-	-	-	Gautamīputra Siri Satakarṇi
3.	24 Savachhare	-	Rainy/ Vasa	-	4 Pakhe	5 Divase	-	-	-	-	Gautamīputra Śātakarṇi
4.	24	-	Summer/ Gimha	-	2 Pakhe	10 Divasa	-	-	-	-	Gautamīputra Śātakarṇi
5.	6 Samvatere	-	Summer/ Gimha	-	-	Panchama Divase	-	-	-	-	Vāśiṣtiputra Śrī Pulāmāyisa
6.	19 Savichhare	-	Summer/ Gimha	-	2 Pakhe	13 Divase	-	-	-	-	The great Queen Gōtamīputra Bālasiri of Śrī Sātakarṇi Gōtamīputra

Donee	Object of the Inscription	Purpose of the Inscription	Corres. Christian Era A.D.	Remarks or References
13	14	15	16	17
-	-	-	-	Nagārjunakoṇḍa Prākṛt Insc. of Gautamīputra V. Satakarṇi, year 6; EI. XXXVI, No. 33, p. 273-274.
To the Tekirasi ascetics	Granted 200 hivartans of land in the village of Western Kakhedi	-	-	Nāsik Insc., EI. Vol. VIII, No. 4, Plate (ii) (ksh. 13).
To the mendicant ascetics of Tiramhu	100 nivartane of land given	-	-	Nāsik Insc.; EI.Vol. VIII, No. 5, Plate (ii) (ksh 14), p. 73-74.
-	the land was given	-	-	Nāsik Insc; EI.Vol. VIII; No.5, Plate (ii) (ksh 14), p. 73-74.
To the sangha of Monks	-	-	-	The Insc. in Nāsik caves No. 1 Plate (iii) ksh 16, EI.Vol. VIII, No. 8, p. 59.
To the sangha of Monks	-	-	-	The Insc. in Nāsik caves No. 8, No. 2, Plate (i) ksh 18, EI.Vol. VIII, No. 8, p. 65-67.

1	2	3	4	5	6	7	8	9	10	11	12
7.	22	-	Gi pa	-	-	7 Diva	-	-	-	-	-
8.	2 Samvachhare	-	Winter/ Hēmanta	-	4 Pakhe	8 Divase	-	-	-	-	Siri Pulūmāvi S/o Vāśiṣṭi
9.	7 year of the king Śri Pulūmāvi	-	Summer/ Gimha	-	5 Pakha	1 Divase	-	-	-	-	Mahārathi Sōmadēva of Okhlakiyas
10.	12 samvach-haram of Harītīputra Śātakarṇi	-	Winter/ Hēmantāna	-	7 Pakha	1 Divase	-	-	-	-	Mahābhuvi (Mahahōji) the king's daughter, Śiva Khaṇḍanāgasiri W/o Jivaputra of a Nāga
11.	8 sava of Śrī Pulūmāvi	-	Winter/ Hēmantāna	2	1 Pakha	1 Diva	-	-	-	-	Gahapatika (house holder) a resident of Vēpūraka Village
12.	24 Savach-har of Siri-Pulūmāvi		Winter/ Hēmantana	-	3 Pakhe	2 Divase	-	-	-	-	Harapharāna a lay worshipper

13	14	15	16	17
-	-	-	-	Nasik Insc., EI.Vol. VIII; No. 25; Plate (vi) (ksh 15), p. 93-94.
A husbands man Dhamana	-	Caused this to be made together with his father and mother with	-	Nasik Insc.; EI. Vol. VIII; No. 25; Plate (vi) (ksh 15), p. 94-95.
To the community of Vāṭūraka	-	-	-	Kārle cave Insc. No. 14; Plate II, (ksh 17); EI., Vol. VII, No. 7; p. 61.
-	A tank and vihara	Meritorious gift of Mahābhuvi	-	Banavāsi Insc. of Vinhukada Śātakanni; year 12; EI, Vol. XXXIV, No. 38, p. 241.
-	To record the sinking of the reservoir	-	-	A New Andhra Insc. or Siri Pulūmāvi, EI.XIV, No. 9, p. 153-155.
To the universal sangha as a special property of the Mahāsāṅghikas	A nine celled hall has been given	in honour of his parents and for securing the welfare and happiness of all beings	-	Kārle cave Insc. No. 20; plate III (ksh 21); EI. VII, p. 71-72.

1	2	3	4	5	6	7	8	9	10	11	12
13./	5 year of some Sove-reign	-	Winter/ Hēmantāna	-	Pakha	2 Divase	-	-	-	-	by nuns
14.	14 sava	-	Rainy/Va	-	4 pa	1 Diva	-	-	-	-	Śivakadaguta (Śivakanda Gupta)
15.	27 Samvach-hara of Gotamiputra Sri Yana Satakarni	-	Winter/ Hēmantāna	-	4 pakha	5 Divase	-	-	-	-	Some Buddhist donation

13	14	15	16	17
-	Commemorated the donation of a cistern made	-	-	Kārle cave Insc. No. 21, Plate IV (k 18); EI. VII, No. 7, p. 73-74.
-	The king secured the property of the village Karajaka	as Monks land	-	Kārle cave Insc., No. 19; Plate II, (k 20); EI. VII, No. 7, p. 64-66.
-	Record of Buddhist Donation	The purpose is not quite certain	-	A New Insc. of the Andhra king Yajña Śrī Gautamīputra; EI. Vol. I, No. XV; p. 95-96.

Regnal Years : The Ikṣvākus (200 A.D. to 260 A.D.)

Sl. No.	Samvatsara	Āyana	Season/ ṚTU	Month/ Māsa	Fortnight/ Pakṣa	Day/ Tithi	Star/ Nakṣatra	Karaṇa	Yōga	Muhūrta	Donor
1	2	3	4	5	6	7	8	9	10	11	12
1.	5 Vijaya Samvatsar of Vāśiṣṭīputra Śāntamūla I	-	Rainy/ Vaṣānām	-	1 Pakha	1 Divasa	-	-	-	-	The Merchant Cūḍādhara Śrī with his wife, daughter, group of daughters-in-law, friends, grandsons
2.	13 of Vāśiṣṭīputra Śāntamūla I	-	Winter/ Hēmanta	-	1 Pakha	1 Divasa	-	-	-	-	The Merchants Mahācandra and Kṣudracandra
3.	Cyclic year "Vijaya" belongs to V.P. Dutta I	-	-	-	-	-	-	-	-	-	-
4.	Cyclic year "Vijaya" belongs to Ehuvala Cāntamūla	-	-	-	-	-	-	-	-	-	-

Donee	Object of the Inscription	Purpose of the Inscription	Corres Christian Era A.D.	Remarks or References
13	14	15	16	17
To a maṇḍapa of the worshipful of the Buddha's sangha	Records a stone pillard Insc.	For his own longevity and continued victory	-	Arch. Series No. 14 Incs. of the Ikṣvākus by P.R. Srinivasan and Sankarnarayanan No.1, p. 7 (pub. in EI. Vol. XXXVII, pp 31 ff)
-	It records the setting up of a pillar called Buddhi-Nikambha in the great chaitya of the monastery of the Bahuśrutīyas	For the unity with their own relatives	-	Ibid, p. 7, No. 2 (pub. in EI.XXXVIII, Epigraphica of Andhrica Vol. I, pp. 146 ff).
-	-	-	273 A.D.	EI.XXXV; JAHRS XXX. Parts I to IV, p. 135.
-	-	-	333 A.D.	EI.XXXV; JAHRS XXX, Parts I to IV, p. 135.

1	2	3	4	5	6	7	8	9	10	11	12
5.	5 Regnal year of King Kumāravara-datta	-	Winter/ Hēmanta	-	4 pakhe	15 Divase	-	-	-	-	Bōddhadatta and Mātrījanapālita, the ministers of the king Vāśiṣṭiputra
6.	6th year	-	Summer/ Gimha	-	6 pakhe	10 Divase	-	-	-	-	-do-
7.	4 Regnal year of Siri Rulu Purisha Dutta	-	Summer/ Gimha	-	8 pakhe	10 Divase	-	-	-	-	Noḍuka-Siri
8.	Samvachhara Vijayam	-	Rainy/ Vasa	-	1 pakha	2 Divasa	-	-	-	-	Sarasika & Bhūsumalata
9.	6 year of V.P. Datta	-	-	-	-	-	-	-	-	-	Śāntiśrī
10.	6 year of V.P. Datta	-	-	-	-	-	-	-	-	-	Śāntiśrī
11.	6 year of V.P. Datta	-	-	-	-	-	-	-	-	-	Śāntiśrī

13	14	15	16	17
The Brāhmaṇas	Records certain pious Donations-1000 cows with two dates.	For the religious merit of the donors parents	-	Gunj Rock Insc. of Kumāravaradatta, EI, XXVII, No. 11, p. 41.
-do-	-do-	-do-	-	Gunj Rock Insc., of Kumāravaradatta, EI. XXVII, No. 11, p. 41.
To the Lord of Hālampura	To record the gift of a field	For securing the longevity of the donor	-	Gurzal Brāhmin Insc; EI.XXVI, p. 123 & 124.
-	Records the erection of gift of memorial pillar of the deceased King Śāntamūla	For the religious merit of the deceased King Śāntamūla	-	Arch. Series No. 14 Inscs. of the Ikṣvākus by P.R. Srinivasan and Sankarnarayanan, No.3, p. 8, (pub. in EI, XXI, p. 63; XXXV, p. 3).
-	Records the donation of pillar	-	-	Inscs. of Ikṣvākus; No. 4. p. 8; EI, XX, p. 14, X.
-	Records the gifts of a pillar	-	-	Arch. Series No. 14 Inscs. of the Ikshvakus; Rajan Veerapurushadatta; No. 5, p. 9 (pub. in EI.XX, p. 14)
-	Donation of a pillar	-	-	Ibid., No. 6, p. 9 (EI, XX, p. 13) Rājan Vīrapuruṣadatta.

1	2	3	4	5	6	7	8	9	10	11	12
12.	6 year of V.P. Datta	-	-	-	-	-	-	-	-	-	Śāntiśrī
13.	6 sava of V.P. Datta	-	-	-	-	-	-	-	-	-	Mahātalavāri Śantiśri, a Mistress of Munificence a Uterine sister of Vīrapuruṣacatta
14.	6 Samva of V.P. Datta	-	Rainy/Va	-	Pa	10 Diva	-	-	-	-	Mahātalvāri, the daughter of the Vāśiṣṭiputra Ikṣvāku Śāntamūla
15.	14th year of Mātarīputra (VP Datta)	-	Winter/ Hēmantāna	-	6 Pakha	13 Divasa	-	-	-	-	Rājan Mātarīputra
16.	15 Samva of M.P. V.P. Datta	-	Rainy/Va	-	-	-	-	-	-	-	Śāntiśrī

13	14	15	16	17
-	Donation of a pillar	-	-	Ibid., No.7, p. 9 (EI, XX, p. 13) Rājan Vīrapuruṣadatta.
-	Records a settng up of the pillar	For attainment of welfare and happiness in both worlds the future and present belonging to her own two families and her own salvation and for the walfare and happiness of the world.	-	Ibid., No. 8, p. 10; (EI.XX, p. 16f) Rājan Vīrapuruṣadatta.
-	setting up of the pillar at the Mahāchaitya of the lord Buddha, the Supreme	Having due record for both the families of her own, for attaining her own welfare and happiness in both the worlds.	-	Ibid., No. 10, p. 11; (EI, XX, p. 18) Rājan Vīrapuruṣadatta
-	A chaitya provided with flooring of stone slabs with a chaitya	For the acceptance of the followers of great teachers & the senior monks	-	Ibid. No. 17, p. 14; (EI, XX, p. 22), Rājan Mātarīputra V.P. Datta.
For the happiness and liberation of the Buddhist monks of different countries	Records the erection of something equipped with all necessaries in the neighbourhood of Mahāchaitya	For the longevity and contd. victory of (her son-in-law) the king M.P. V.P. Datta	-	Ibid., No. 18, p. 15; (EI. XXI, p. 66, 67; XXXVIII, p. 175-177).

1	2	3	4	5	6	7	8	9	10	11	12
17.	15 Samva of M.P. V.P. Datta	-	Rainy/Va	-	-	-	-	-	-	-	Śāntiśrī
18.	18 Samvach-Charama of VP Datta	-	Winter/ Hemantana	-	6 Pakha	15 Divasa	-	-	-	-	Mahātalvāri Śāntiśrī
19.	19 Samvach-charam of V.P. Datta	-	Summer/ Gimha	-	4 Pakha	13 Diva	-	-	-	-	Samghila, the son of the Merchant Vaira
20.	19 Samva-chara of M.P.V.P. Datta	-	Rainy/ vasa	-	8 Pakha	10 Diva	-	-	-	-	Siddhartha, son of Nāga candra, an artisan

13	14	15	16	17
For the acceptance of the Apara-mahāvinayasailiya	Records the erection of something with a quadrangular building at the Mahāvihāra near the Mahāchaitya	For the welfare, happiness in both the worlds and for her liberation and for a 100 years of life of V.P. Datta	-	Ibid., No. 19, p. 15; (EI. XXI, p. 66) VP. Datta.
-	Records the erection of a stone hall surrounded by and provided with all facilities	For the longevity and victory of her son-in-law M.P. V.P. Datta and for her own welfare and happiness of the both the worlds and for her liberation having due regard to the highly miserable of the past, future and present times of her own two families.	-	Ibid., No. 20, p. 16; (EI. XX, p. 21) VP. Datta.
-	Records the registering of construction of something at a Mahāvihāra situated in a Sea-port town.	For the merit of his mother, wife, kinsmen, friends, relatives etc.	-	Ibid., No. 21, p. 16, 17; (EI, XXXIII, p. 191), VP Datta.
-	Records the crection of 5 entrance pillars with all accessories at the eastern enterance of the Great Caitya if the Bhāgavat. Buddha in the vill. Vēlagiri	-	-	Ibid., No. 22, p. 17; (pub. in A.S.S.I., Vol. I, p. 110.)

SUMMARY

The various schools of Indian thought have diversified notions of Time. There is mention of Time in the *Vēdas*, *Upaniṣads* and *Purāṇas*. One important characteristic of this period is to treat Time as to be fundamental and very important in the process of evolution. Sometimes Time is made to occupy a place above everything else, even above God.

In the earliest days of human thinking in India, the discussion about Time was guided or determined by certain myths, which were considered as temporal and which were prevalent among the people. The significance of myths lies in the fact that they have always played a very important role in shaping the various conceptions of reality in our cultural and philosophical terms. They have thus proved to be in many cases, indicative of truth.

In the *Atharva Vēda,* Time is regarded as the Supreme Ontological Being, Cosmic Power, Principle of casualty, and as Supreme Divinity, in contra-distinction of the classificatory being "when everything in the Universe is classified as Being, and is placed under Being as the highest class." To this 'Time', even the supreme creator God, Prajāpati is subordinate. Time is thus metaphorically considered as the Creator of the Creator. In other words, even the Creator is and has His Being in Time. This means that whatever has a history, anything other than the Cosmic Principle as oneness, is temporal and is talked in the framework of time. Even the Cosmic Principle as eternal duration is understood with reference to Time. This is actually the view of those who reflect on Time as the cause of everything, every classificatory being of the Universe. These thinkers are technically called *Kālacintakas, Kālavādins* or *Kālakāraṇavādins*.

The purpose of placing Time above even the Supreme God is to remind of him of his temporality, and to show that he is not the Supreme Reality. This enlightens him to rise above his vanity and makes Him realize that Supreme Reality is something beyond. This is evident from the contexts in which some of the myths are narrated to Gods like Indra. Further, such characterization of Time is a factor in revealing the Ultimate Reality to Gods as well as Human Beings. Such myths also terrorize them to ponder over their temporal and implicitly evanescent existence, so that they pursue the highest value.

The word used for Time, in Indian Philosophy, is *Kāla* derived from the root *'Kal'* meaning to *'count'* or 'to *devour*'. Thus Time is that which is the basis for counting temporal units, such as Seconds, Hours, Days etc., or it is that which devours all its beings as its food. In other words, Time measures or scales every being from its birth to death. *Kāla* is so called because it, like the cyclical activities and dynamism of a water-wheel and being all-pervasive, determines or accounts for the birth, subsistence and destruction of all empirical beings. Because of this function of Time, which resembles the function of the Creator God, the *Kṛṣṇa* in the *Gīta* has identified himself with Time.

The myths of Time in different forms, sometimes personified as Superme Deity and sometimes with various Gods in their different roles are found in the *Vēdas, Upaniṣ ads, Mahābhārata, Bhagavad Gīta* and the *Purāṇas* and meant to serve a cultural and philosophical purpose.

One of the functions of the myths is to describe the

efficacy of Time, and to create images. In ancient times, rituals and sacrifices *(Karma)* were performed in order to regenerate Time by a series of rituals, which in a sense re-actualize the Cosmogony. Thus, *Kāla* and *Karma* were considered to be the two Supreme forces of the Universe.

The discussion of the myth of Time in the *Brahmavaivarta Purāṇa* elucidates its function in destroying God *Indra's* vanity of pride of being in an exalted position. It makes him realize his transitory and temporal character and the falsity of his pride and position. Indra feels enlightened, transcends his empirical status and surrenders to the mythical sacred Time, which is equivalent to a revelation of the Ultimate Reality. And this is a strictly Metaphysical Reality, accessible only through myths and symbols. There is another important myth of Time characterized as a rotating and dynamic wheel i.e., *Cakra*. This symbolizes the cycle of the Universe and human existence and shows that every empirical being is under the wheel of Time for its birth, subsistence and death.

Time is very often signified as Death. The Cosmological role of *Kāla* is frequently described in the *Purāṇas,* which says that Time exists in a latent state during the dissolution of the world. And *Kāla* is closely linked to *Śiva;* it is its destructive aspect, which dominates all other Cosmological aspects. *Śiva* himself is called *Mahākāla*, Great Time, meaning Death. The destructive Goddess *Kāḷi* is the female counterpart of the God *Rudra-Śiva,* who is identical with Time. Such signification of myths influences, at least, human being who then tries to seek refuge under the Higher Reality than Time itself. This search leads him to the realization of or even his identification with the Supreme or Ultimate Principle of the Universe. Consequently, he transforms his total personality and attitude towards his fellow beings. His behaviour *(Karma)*-mental, physical and linguistic is guided accordingly. His good *karmas,* it is maintained, get him rid of the infinitely recurring empirical existences, which are intrinsically and ultimately painful in nature. The realization of the Ultimate Reality and the Law of *Karma,* in respect of a person, have religious, cultural and social relevance.

One of the most popular images of Time is that of a river. Time is ever flowing like a river. It carries away beings from one state to another and causes changes in their qualities.

In the Post-*Vēdic* literature, such as the *Upaniṣads* and *Vēdānta* texts, we find a radical departure in thinking about time. In the *Vēdas,* the philosophical trend was not very clear, but in the *Upaniṣads* it became monistic, which further took the form of Absolutism. With this change, the status of Time as Supreme Reality also changed. Now it is considered to have a derivative status following the changing phenomena. It was thus treated, as subordinate to appearance not related in any sense to Ultimate Reality, which is Timeless. This idea is very vaguely found in the *Atharva Vēda* (XIX, 53.3.) where Time is said to flow from the vessel of Ultimate Dynamic Reality *(Brahman)* which in its unchanging form is Timeless, and in its manifest form i.e., as the course of the phenomena is Time. In other words, the temporality of changing phenomena gives rise to the notion of Time, the experience of which is gained through the spatial periodic moments of the celestial bodies like Sun.

At phenomenal level, Time is metaphorically said to cook or mature or devour all beings. Here it is Time which accounts for the birth, sustenance and death of every being. That is, as long as there is change, there is need of time to account for it. But when the Ultimate Reality is in its original primitive state, there is no change and phenomena and hence, no experience or need of Time. In such a reality, Time itself is devoured. This leads to the unreality of Time in ultimate sense. In other words, Time is the essence of the phenomenal world. And since the phenomenal world has no independent reality, Time too is not an independent reality.

The source of time-consciousness and of Time is apperception, which is not something different and separate from the 'I am'. It has no being of its own, when not activated by the I am. The present then of Time is the presence of the aspect of I am in apperception which collects together past and future into the present. This collecting is done by the memory and anticipation of the I am in apperception. This way of Time-consciousness, in terms of past, present and future is a mental event or derived from Psychology having reference to consciousness. It is based on the psychology of remembering, perceiving and anticipating. This is called psychological reduction. Time is thus subjective in the sense that these psychological states take a subject, and wherever there is subject, there is subjectivity. This subject has in its reflective consciousness the awareness of I am through which these states are interrelated, that is, they are given an order which further requires Time to be explained. If Time and temporality Past, Present and Future are psychological, then temporal or phenomenal becoming is also psychological. The temporal becoming of phenomenal events can thus be reduced to mental events. The *Maitrāyaṇi-Upaniṣad* mentions Time as one of the principal forms of the supreme, immortal, unembodied *Brahman.* It is as a form of *Brahman* that Time is to be meditated on. Therefore, meditation on Time is recommended as a method of getting beyond Time to the Timeless Reality.

The concepts of Times change and causality are related to each other in such a way that one requires the other to be understood, but change occupies the central place among them. Epistemologically, change precedes Time, but it requires time to be elucidated. The notion of causality arises when there is change in a thing from one event to another, the preceding event as cause and the succeeding event as effect. Now Time is brought into account for this change and causality. But Time as a principle of elucidation is understood only when there is change with casual continuity. The main difference between Change and Time is that whereas change enters direcly into our experience, time does not. The concept of Time is actually derived from change. The same is the case with the concept of Causality. In philosophical systems of India, the metaphysical position of a system determines the theories of change and causality, which in turn determine the theory of Time. But in all philosophical systems of India, even in *Buddhism,* the underlying essence maintains its identity throughout changes, which are taken as its passing phases or modes.

The *Nyāya-Vaiśēṣika* realism ascribes reality to these modes also, and subscribes to the views of a theory of new creation *(Ārambhavāda)* which is their theory of causation.

For *Sāṅkhya-Yōga,* a mode is only a state of the unchanging, underlying and identical essence. It is entailed in its substratum before its coming into existence. Its reality is more epistemological and ontological. The *Sāṅkhya-Yōga* did not exploit the full implications of their theory of Causation. In some sense, they also subscribe to the reality of these modes, together with the reality of difference and change.

Further both *Nyāya-Vaiśēṣika* and *Jainism* also talk of duration. They form the concept of duration on the basis of their experience of perceptual or spacious present, which ranges from one moment to a number of moments. Their notion of present is exemplified in the subsistence of an object, on a course of action like 'cooking' including its whole process. Then they conceive of the absolute and substantive Time by extending the spacious present in forward as well as backward direction to its extremity on both sides. They also conceive of Absolute Time as a substratum of the temporal relations, temporal properties and temporal acts, which we use in our ordinary language and which, we have found to be psychological in origin. This shows that they admit Time as a substantive, and give its descriptions via temporal properties.

Nyāya and *Jainism* have also discussed the nature and duration of present Time in some detail. Both maintain two kinds of Time-one, which is absolute and substantive *(Mahākāla, Paramārthakāla, Dravya Kāla, Niścaya Kāla)* and the other is the empirical divisions of Time *(Khaṇḍa Kāla, Vyavahāra Kāla, Samaya)* based on spatial periodic movements of physical bodies like the Earth and the hands of a watch. Both infer Time, on priority grounds, as the underlying explanatory principle of all changes, events, happenings, processes and subsistene *(Sthiti, Vartana, Dhrauvya)* of their substances in their unchanging essential forms. In these systems, the a priori notion of Absolute Time is derived from the a priori notion of present, which is inseparably related, and parallel to the eternal essence of other substances. They also subscribe to empirical notion of present, which is either extensionless or extended. The extensionless present is logically instantaneous and durationless, while the extended present is more psychological, and has duration of infinite dimension depending on the context.

In the *Jaina* philosophy *Kāla* is as much a Real Substance as the five others, viz., *Jīva, Dharma, Adharma, Pudgala, Ākāśa*. It is described as the accompanying cause or condition *(Sahakāri Kāraṇa)* or *Asamavāyi Kāraṇa,* as the *vaiśēṣikas* call it, of the modification of Substances. Yet Time is a real factor in the happening of their modifications. In Jain philosophy, a real thing is said to be characterised by its three aspects respectively called *Utpāda* or Origination, *Vyaya* or Annihilation and *Dhrauvya* or Persistence. Time in its impermanent aspects, viz., the aspects of origination and decay-*Utpāda* and *Vyaya* is called the *Paryāya Kāla* (Time in modification) or *Vyavahāra Kāla* (Time in ordinary practice). It is technically called *Samaya*.

In Buddhism, the Time concept evolved from the experience of the fact of change, which led to accepting it as the basic assumption of the Buddhist schools. Early Buddhism considered both Time and Causation as parts of our experience, not as more inferences based primarily on the succession of momentary ideas. Buddha seems to have considered Time as an essential feature of the Universe and the experience of it. The Buddha's statement regarding Time should be understood not at its face value, rather through the interpretation of the context in which he held an empiricist view of Time. In due course after the Buddha's passing away the empiricist view of time and temporality changed completely. Such shift was a logical implication of the Buddha's certain statements about the nature of the Universe. The Buddhists would have no problem at all in accepting or incorporating the conventional way in which we speak of Time, that is in terms of the three temporal periods past, present and future. It is the function of the mind, after all to conceive of time in that order or in deed to give order to the nature of things. Thus, whether it is simple clock time, physical time (measurements of movements) or psychological time, the mind knows or senses time because of the abstractive quality. A Buddhist will never talk of Time other than temporality.

The Cyclic Time occurs frequently in cosmological context such as Hindu theory of *Yugas, Manvantaras* and *Kalpas,* whereas Linear Time in historical sources such as

Genealogies, Biographies and Chronicles, where Time reckoning was recorded through Generations, Regnal Years and Eras. Cosmological Time relates to the Universe is almost infinite and where it is cyclic, consisting constructions of many cycles of Time setting out as it were a Calendar for the Universe. A more realistic kind of the time reckoning can be drawn on actual historical needs. The use of Eras as well as dating by Regnal Years had gained currency, simultaneously with activity in Astronomy and Cosmology.

The natural divisions of Time-Years, Months, Days have in all ages, being determined by the motions of the Sun and the Moon. The special occasion for gifts amply borne in Epigraphical records signify the application of Time and its importance.

Tabular statements analysing historical insciptions up to the seventh century A.D. include *Vikrama, Śaka, Gupta, Harṣa, Vallabhi, Cēdi, Kālacuri* Eras and *Regnal* Year reckoning.

BIBLIOGRAPHY

PRIMARY SOURCES

Brahmanical Works

Atharva Vēda Saṁhitā (Śaunaka) with Sāyana Bhāṣya : Ed. by Viśva Bandu, Hoshiarpur, 1960-64.

Jaiminīya Saṃhitā : Ed. W. Caland, Breslau, 1907.

Kāṭhaka Saṃhitā : Ed. S.D. Śatāvalēkara, Audh, 1943.

Ṛgvēda Saṃhitā with Sāyana Bhāṣya : Ed. N.S. Sōntakkē and C.G. Kāṣīkar, Poona, 1933-31.

Sāmavēda Saṃhitā : Ed. S.D. Śatāvalēkara, Pardi, 1956.

Taittirīya Saṃhitā : Ed. S.D. Śatāvalēkara, Pardi, 1957.

Aitarēya Brāhmaṇa with Sāyana Bhāṣya : Pub. in An. S.S., Poona, 1930-31.

Ārṣēya Brāhmaṇa with Sāyana Bhāṣya : Ed. B.R. Śarma, Tirupati, 1967.

The Atharva Vēda and The Gōpatha Brāhmaṇa : Bloom Field. M. (Grundris II, 1 B), Strassburg, 1899.

Gōpatha Brāhmaṇa : Ed. D. Gaastra, Leiden, 1919.

Jaiminīya Brāhmaṇa : Ed. Raghuvīra and Lōkēṣ Candra, Nagpur, 1954.

Āpastambhīya-Gṛhya Sūtra : Ed. M. Winternitz, Vienna, 1887.

Āśvalāyana Gṛhya Sūtra : Ed. V.S.S. Ranade in An. S.S. Poona, 1936.

Bauddhāyana Dharma Sūtra : Ed. Dr. E. Hultsch, Leipzig, 1882.

Gōbhila Gṛhya Sūtra : Ed. Cintāmaṇi Bhaṭṭācārya, Calcutta, 1936.

Jaiminīya Gṛhya Sūtra : Ed. & Tr. Caland, Lahor, 1922.

Parāśkara Gṛhya Sūtra : Tr. Olden Berg, In SBE, Vol. XXIX, Oxford, 1886.

The Mantrapāṭha of the Āpastambhins : Ed. M. Winternitz, Oxford, 1897.

Agni Purāṇa : Tr. Manmathanāth Dutt Śāstri, 2nd edition, Vāraṇāśi Chowkamba Sanskrit Series office, 1967.

Bhāgavata Purāṇa : Gaṇapati Kṛṣṇājī Press, Bombay, 1889.

Bhāgavata Purāṇa : (Mythological Series special study) Sindhu. S. Dange, Ajanta Publications, Delhi, 1984.

Bhāgavata Purāṇa : Tr. Ganesh Vasudeo Tagore, Motilal Banarsidass, (Vol. VII-XI part 1-5), Delhi, 1976.

Brahmāṇḍa Purāṇa : Sri Venkateswara Press, Bombay.

Brahmāṇḍa Purāṇa : Tr. Ganesh Vasudeo Tagore, Motilal Banarsidas, (Vol. XXII pt. 1).

Garuḍa Purāṇa : Ed. J.L. Shastri, Motilal Banarsidas, Delhi, 1978.
(Vol. XII, XIII, XIV pts. 1, 2, 3).

Liṅga Purāṇa : Ed. J.L. Shastri, Motilal Banarsidas, Delhi, 1973.

Mārkaṇḍēya Purāṇa : Tr. F. Eden Pargiter, Indological Book House, Delhi, 1969.

Matsya Purāṇa : Anandasrama Sanskrit Series, Poona, (revised 1907).

Sādhanamāla : Bhattacharya. B. Gaekwad, Oriental Series, (2 Vols.).

Śatapatha Brāhmaṇa : *'Sacred Books of the East'* Tr. Julius Eggeling, (5 Vols) Nos. 12 (1882), 26 (1885), 41 (1894), 43 (1897) and 44 (1900), London.

Taittirīya Brāhmaṇa Sāyana Bhāṣya : Pub. in An. S.S., Poona, 1934-38.

Tāṇḍya Mahā-Brāhmaṇa : Ed. A. Chinnaswamy Sastry, Varanasi, 1935.

Aitarēya Upaniṣad : Ed. S. Radhakrishnan in the Vol. entitled 'The Principal Upaniṣads', London, 1953.

Bṛhadāraṇyaka Upaniṣad : Included in S. Radhakrishnan's 'The Principal Upaniṣads'.

Śvetāśvatara Upaniṣad : Included in S. Radha Krishnan's "The Principal Upaniṣads'.

Taittirīya Upaniṣad : Included in S. Radhakrishnan's 'The Principal Upaniṣads'.

Vāyu Purāṇa : Anandasrama Sanskrit Series, Poona, 1905.

Vēdic Index : Macdonnel & Keith, 2 Vols., London, 1912.

Viṣṇu Purāṇa : A System of Hindu Mythology and Tradition: Pub. Sankar Battacharya for Purthi Pustak, Calcutta, 1972.

Viṣṇu Purāṇa : Jīvānanda Vidyāsāgar, Calcutta, 1882.

Viṣṇu Purāṇa : Tr. H.H. Wilson, London, 1864-70. (5 Vols.), Reprint-Calcutta, 1961.

Viṣṇu Dharmōttara Purāṇa : Tr. H.H. Wilson, London.

Jaina & Buddhist Works

Anguṭṭara Nikāya : PTS, E. Morries and E. Hardy. London, 1885-1900, (5 Vols), Tr. F.L. Woodward and E.M. Hare. The Book of the Gradual Sayings, *PTS. 5 Vols. London, 1932-36.*

Dammapatha : Ed. S. Sumangala Thera, London, 1914. Tr. C.A.F. Rhys Davids. PTS. 1931.

Gommaṭasāra Jīva Kāṇḍa : Shri Nemichandra Siddhanta Chakravarti, Ed. With introduction, Trans and commentary by Rai Bahadur J.L. Jaini, Pub. Pandit Ajit Prasada, SBJ, Vol. 5, The Central Jaina Pub. House, Ajitashram, Lucknow, 1927.

Kāthāvattu : Ed. Taylor, 2 Vols., 1894-97, Tr. Points of Controversy. PTS 1915.

Lalita Vistāra : Ed. P.L. Vaidya, Darbhanga.

Pañcāstikāya : Kuṇdakunda, SBJ.

Prapaṇcasāra : Kundakunda, Sholapur, 1947.

Suttanipāta : Ed. D. Anderson & H. Smith, London, 1948. Tr. V. Fausboll, SBE. Vol. 10, Part 2, Oxford, 1881.

The Jātakas or Stories: Ed. Cowell, E.B. 1895 (6 Vols),

of the Buddha's former Births Clarandon Press.

The Tatvasaṅgraḥ of 'Śāntarakṣita' : With the commentary of Kamalasila, Oriental Institute, Baroda, 1939, Vol. II, pp. 861-86.

Vinaya Pīṭaka : H. Oldenberg, 5 Vols. London, 1879-83 Tr. T.W. Rhys Davids & H. Oldenberg, The Book of Discipline, SBE, Vols. 13, 17, 20, Oxford, 1881-85.

Other Texts

Āśvalāyana Gṛhya Pariśiṣṭa : Ed. P.K. Aithal, Adyar Library Bulletin, Madras, Vol. XXVII.

Bāsvati : Satananda, Ed. Matri Prasad Pandey (Kashi Sanskrit Series)., Banaras, 1917.

Bhagavadgīta : Pub. by Gita Press, Gorakhpur.

Kāmasūtra of Vātsāyana : Ed. Gosvami Damodar Sastri, Kashi Sanskrit Series, Varanasi, 1929.

Kauṭilya-Artha Śāstra : Ed. R. Shamasastry, Mysore, 2nd edition, 1919.

Kauṭilya Artha Śāstra : Kangle, R.P. (Ed & Tr), 3 Vols., Bombay, pt. I (text), 1960; pt. II (translation), 1963, pt. III (critical study), 1965.

Mahābhārata : With the commentary of Nīlakaṇṭh Ed. Ramachandra Śāstri Kinjwadekar, Poona, 1929-33.

Manusmṛti : English Translation by G. Buhler in S.B.E. Vol. XXVI, Oxford, 1886.

The Nighaṃṭu and Nirukta : Ed. L. Sarup, Lahore, 1927.

Nirṇaya Sindhu : Kamalakara Bhatta, Telugu translation by Dr. Kidambi Narasimha Charya, Part-I, Hyderabad, 1995.

Rāmāyaṇa of Vālmīki: Ed. R. Narayana Swami Aiyar, Madras, 1933.

Smṛti Candrika : Ed. L. Srinivasacharya, Mysore, 1914.

Śrīmad Bhāgavatam : Bhakti Vēdānta, Part-III, Pub. by The League of Devotees (Reg), Vrindavan, Swami, A.C., P.B. No. 1846, Delhi-6.

Śrīman Mahā-Bhāratam : Ed. Krishnacharya, T.R. & Vyasacharya, T.R. according to Southern recension based on the South Indian texts with footnotes and readings. Śāntiparva-XII, Sri Sadguru Pub., Delhi, 1985.

Ten Principal Upaniṣads : Works of Śaṅkarācārya in original with Śaṅkara Bhāṣya Sanskrit, Vol. I, Pub. Motilal Banarsidass, Delhi-7.

Vēdaṅga Jyōtiṣa : Ed. With his own Sanskrit commentary and English translation by R. Shama Sastry, Mysore, 1936.

Viṣṇu Dharmasūtra : Jolly, J. (Tr) 'Sacred Books of the East', Vol. VII.

Yājñavalkya Smṛti : With Mitākṣara commentary of Vijñānēśvara, Pub. Nirnaya Sagara Press, Bombay, 1926.

Yajña-Tatva-Prakāśa : Sastri A Chinnaswami, Madras, 1953.

Bloom Field, M. · *'The Hymns of the Atharvavēda'*, S.B.E. Vol. XLII, Oxford, 1897.

Max Muller, F : *'Sacred Books of the East'* 50 Vols., Oxford, 1979-1900.

Monier Williams : *'Brāhmanism and Hinduism'*, John Murray, London, 1891, (4th Edition).

Sen, Chitrabandhu : *'A Dictionary of Vēdic Rituals'*, Delhi, 1978.

Swamikannu Pillai, L.D. : *'An Indian Ephemeris (AD 700-AD 1799)'*, Vol. I, Part-I, Pub. Agam Prakasan, Delhi, 1982.

Vikrata Rama Chandra : 'Vidyāraṇya Pañcānan, Mānāntara Bōdhan' (Date calculator), Vijaya Press, Acharyulu Warangal, 1971.

Vishva Bandu : *'Vaidika-Padānukrama-Kōśa'*, Vols. I-XVI, Lahore & Hosiarpur, 1935-65.

Archaeology & Epigraphy

Annual Reports of the Archaeological Survey, Mysore (A.S.M).

Annual Reports of the Superintendent for Epigraphy, Southern Circle, Madras. From 1875-1930 (A.R.S.I.E.).

Annual Reports of the Archaeological Survey of India (A.S.I.), for the years 1906-07 & 1926-27 to 1934-35.

Annual Reports of Archaeological Survey, Southern Circle, for the years 1910-1918, Madras (A.S.S.I.).

Annual Reports of the Archaeological Survey of India, Western Circle (A.S.W.I.).

A.S.I.: *Annual Reports of the Archaeological Survey of India,* New Delhi.

A.S.S.I.: *Archaeological Survey of Southern India,* Madras.

Butterworth, Allen & Venu Gopal Chetty. V-*Collection of the Inscriptions on Copper Plate and Stone in the Nellore District (N.I.),* Three Volumes, Madras. 1905 (Arch. Surv. S. India.)

Copper Plate Inscriptions of the State Museum. Vol. II, by Sri N. Ramesan, Pub. by The Govt. of A.P., Hyderabad, 1970.

Corpus of Inscriptionum Indicarum, 4 Vols., London and Ootacamund, 1888-1955.

Corpus Inscriptionum Indicarum, Vol. III, 'Inscriptions of the Early Gupta kings and their Successors', by John / Faithful Fleet, Indological Book House, Bansphatak, Varanasi, 1963.

Corpus of Telangana Inscriptions (Part IV). by Mallampalli Sōmaśēkhara Śarma, Pub. by the Govt. of A.P. Hyderabad, 1973.

Cousens, Henry : *'List of Antiquarian Remains in C.P. and Berars'* (N.I. Vol. IV) (Arch. Surv. India), Calcutta, 1897.

Date of Ashoka's Rock Edicts, The Indian Antiquary, Feb. 1927, page 27.

Epigraphia Indica, Vol. I to XXV and XXVI, New Delhi (E.I.)

Epigraphia Indica and Record of the Archaeological Survey of India, Ed. By Hirananda Sastry, Manager of Publications, Delhi, 1933.

Fleet. J.F. : *'Pali, Sanskrit and old Canarese Inscriptions from the Bombay Presidency and part of the Madras Presidency',* London, 1898.

Hiralal, Rai Bahadur: *'Descriptive List of Inscriptions of C.P. and Berar'*, Nagpur, 1916.

The Indian Astronomical Ephemeris for the year 1982, Delhi, 1981.

Inscriptional Glossary of Andhra Pradesh by K. Iswara Dutt, A.P. Sāhitya Academy, Kalā Bhavan, Hyderabad, 1967.

Inscriptions of the Ikṣvāku Period by P.R. Srinivasan and Śaṅkara Nārāyanan, Pub. by the Govt. of A.P., Hyderabad, 1979.

Luders, H. : *'A list of Brāhmī Inscriptions from the Earliest Times to about 400 A.D.'*, Edited in Epigraphia Indica, Vol. X (Appendix).

Rice, Lewis : *Mysore Inscriptions (Trans. for the Govt.),* Bangalore, 1879.

Select Incriptions, Vol. I by D.C. Sircar, 2nd Edition. S.I.I. *(South Indian Inscriptions)*.

South Indian Inscriptions (S.I.I) Vols. I-VIII. New Imperial Series, New Delhi (Arch. Surv. Ind) : Subrahmanyam. R: *'Tundi Copper Plate Grant of Viṣṇukuṇḍin King Vikramēndra Varman'*, A.P. Govt., Archaeological Series No. 8, Pub. by the Govt. of A.P., Hyderabad, 1962.

Secondary Sources

Ajay Mitra Shastri : *'Early History of the Deccan Problems and Perspectives'*, Sandeep Prakasan, Delhi.

Albert Schweitzer : *'Indian Thought and its Development'*, Hodder and Stonghton Ltd., London, 1936.

Barnett, L.D. : *'Antiquities of India'*, London, 1913.

Barth, A. : *'The Religions of India'*, London, 1914.

Basham, A.L. : *'The Wonder That was India'*, Canberra, 2nd Edition, 1966.

Bhandarkar, R.G. : *'Early History of the Deccan'*

'A Peep into the Early History of India'

'Vaiṣṇavism, Śaivism and Minor Religious Systems', Encyclopaedia of Aryan Research. Vol. 1, III, Strassburg, 1913.

Bloom Field, M. : *'The Religion of the Vēda'*, New York, 1908.

Colling Wood, R.G. : *'The Idea of History'*, Oxford University Press, 1946.

Das, A.C. : *'Ṛgvēdic Culture'*, Calcutta, 1925.

Das Gupta, S.N. : *'History of Indian Philosophy'*, 4 Vols., Cambridge, 1923-49.

Dubois, A.J.A. : *'Hindu Manners, Customs and Ceremonies'*, Tr, into English by H.K. Beauchamp, Oxford, Third Edition, 1906.

Eliot, C. : *'Hinduism and Buddhism'*, (3 Vols.), London, 1922.

Elizabeth Achelis : *'The Calendar For The Modern Age'*, Thomas Nelson and Sons, New York, 1959.

Fergussion, James : *'Tree and Serpent Worship'* Illustrations of Mythology and Art in India from the first Century B.C. to the fourth Century A.D. from the sculptures of Buddhist topes of Sanchi and Amarāvati with introduction essay and descriptions, 1868.

Filita Bharucha : *'Role of Space-Time in Jaina's Syādvada & Quantum Theory'*, Delhi, 1993.

Hopkins, E.W. : *'Epic Mythology'*, Strassburg, 1915.

'Great Epic of India', New York, 1901.

Jayaswal, Kashi Prasad : *'History of India from 150 A.D.* to 350 A.D., Motilal Banarsidas & Co., Lahore, 1933.

Jha, G.N. (Tr) : *'Examination of the doctrine of Traikālya things continuing to exist during Three points of Time'.*

Kane, P.V. : *'History of Dharmaśāstra'*, 6 Vols., Poona, 1930-62.

Kapila Vatsyayan : *'Concepts of Time Ancient and Modern'*, Pub. Indira Gandhi National Centre for the Arts, Central Vista Press, Janpath, New Delhi-1, 1991.

Kashikar, C.G. : *'A Survey of the Śrautasūtras'*, Bombay, 1968.

Keith, A.B. : *'The Religion and Philosophy of the Vēda and Upaniṣads'*, (32 Vols), Harward Oriental Series, 1925.

: *'The History of Sanskrit Literature'*.

: *'The Doctrine of Reality (The Nature of Time)'*, Buddhist Philosophy in India and Ceylon, Oxford University Press, 1923, pp. 163-68.

Kohli, S.K. : *'Time is God'*, New Delhi, 1987.

Kosambi, D.D. : 'The Culture and Civilization of Ancient India', London, 1965.

Krishna Rao, B.V. : *'Early Dynasties of Andhra Dēśa'*, Rajahmundry, 1942.

Krishna Swami Aiyengar : *'Beginnings of South Indian History,'* Madras.

Kumaraswamy, A.K. : *'Time and Eternity'*, Select Books Publishers, Banglore, 1990.

Macdonell, A.A. : *'Vēdic Mythology'*, Delhi, 1980, (Rept).

: *'Sanskrit Literature'*, London, 1900.

Madhavatirtha : *'The Concept of Time in Indian Philosophy, Vēdānta Āśram,* Ahmadabad, 1951.

Maha Lingam, N. : *'Ancient World (A New Look)'*, Pub. International Society for the Investigation of Ancient Civilisation, Poes Garden, Madras, 1983, (Ist edition).

Majumdar, R.C. : *'Ancient India'*, Banaras, 1952.

Majumdar & Others : *'Advanced History of India'*, 2nd edition, London, 1950.

Majumdar & Others : *'History and Culture of the Indian People'*.

Vol. 1-*The Vēdic Age'*, London, 1950;

Vol. II-*The Age of Imperial Unity*, Bombay, 1951;

Vol. III-*The Classical Age*, Bombay, 1954.

Meyer, J.J. : *'Sexual Life in Ancient India'*, 2nd edition, London, 1952.

Mukerjee, S. : *The Jaina Philosophy of Non-Absolutism,* Calcutta, 1944.

Murti, T.R.V. : *The Central Philosophy of Buddhism,* 1955

Nakamura, Hajime : *'Time in Indian and Japanese Thought'.*
'The Voices of Time', Ed. J.T. Froger, New York, G.Braziller, 1966, pp. 77-85.

Nilakanta Sastry, K.A. : *'History of India',* Part-I, Madras, 1950.

Nohara and Ghosa : *'Epitome of Jainism',* Calcutta, 1917.

Padmarajah, Y.J. : *'Jaina Theory of Reality and Knowledge',* Bombay, 1963.

Pandey, R.B. : *'Hindu Saṃskāras'.* Delhi, 1982. (Rept).

Pargiter, F.E. : *'The Purāṇa Text of the Dynasties of the Kali Age',* Oxford University Press, 1913.

Pires, Edward, A. : *'The Maukharis',* B.G. Paul & Co., Madras, 1934.

Radhakrishnan, S. : *'The Hindu View of Life',* London, 1927.

Radha Krishnan. S. : *'Indian Philosophy',* 2 Vols., London, 1923-27.

Ranade, R.D. : *'A Constructive Survey of the Upaniṣadic Philosophy',* Oriental Book Agency, 15, Shukrawar, Poona-2, 1926.

Rapson, E.J. : *'Cambridge History of India',* Vol. I, Cambridge University.

Rhys Davids, T.W. : *'Buddhism, its History and Literature',* 2nd edition, London, 1926.

Romila Thapar : *'Time as a Metaphor of History:Early India'*, Oxford India Paper Backs.

Rockhill, W.W. : *'The Life of Buddha'*, London, 1884.

Satishchandra Vidya-Bhushana : *'A History of Indian Logic, Ancient, Medieval and Modern Schools'*, Readership Lectures, Calcutta University, 1921.

Schayer Stanislaw : *'Contributions to the problem of Time in Indian Philosophy'*, Krakow, 1938.

Schubring, W. : *'The Doctrine of the Jainas'*, Delhi, 1962.

Shah, C.J. : *'Jainism in North India (800 B.C.-A.D. 526)'*, London, 1932.

Smith, V.A. : *'Early History of India'*, 3rd and 4th editions, Oxford.

Stevenson, S. (Mrs.) : *'The Heart of Jainism'*, Oxford, 1915.

Subbaraya Sarma : *'The Greatness of Gāyatri-Jyōtiṣa Śāstra (Origin of Astrology and Astronomy)'*, Bangalore, 1971.

Subrahmanyam, N. : *'Historiography'*, Koodal Publications, Madhurai, T.N., 1973.

Subramaniam, K.R. : *'Buddhist remains of Andhra and Andhra History'*, Andhra University, Waltair.

Swami Madhava Tirtha : *'The Concept of Time in Indian Philosophy', Vēdānta Āśrama,* Ahmedabad, 1951.

Swamy Nikhilananda : *'Self Knowledge', Madras, 'Śrī Rāmakṛṣṇa* Maṭh, 1947.

Tatiya, N.M. : *'Studies in Jaina Philosophy'*, Banaras, 1951.

Thibaut, G. : *'Indischa Astronomie, Astrologie and Mathematik'*, Strassburg, 1899.

Tilak, B.G. : *'The Orion or Researches into the Antiquity of the Vēdas'*, 4th Edition, Poona, 1955.

Vasudeva, S. Agarwala : *'Matsya Purāṇa-A Study (An exposition of the Ancient Purāṇa Vidya)'*, Kanshiraj Trust, Varanasi, 1963.

Venkateswar, V. : *'Indian Culture Through The Ages'*, Vol. I and II, Longmans, London, Bombay, 1928.

Whitrow, G.J. : *'Time in History, The Evolution of our Time and Temporal Perspective'*, Oxford University Press, New York, 1988.

Articles

Bareau Andre : *'The Notion of Time in Early Buddhism'*, East and West 7, 1957, pp. 353-64.

Bedekar, V.M. : *'The Doctrine of Svabhāva and Kāla in the Mahābhārata and other old Sanskrit works'*, Journal of the University of Poona (Humanities) No. 13. 1961.

Bhattacharya, Harisatya : *'The Theory of Time in Jaina Philosophy'*, The proceedings of the second session of the Indian Philosophical Congress, 1926.

Bhattacharya, U.C. : *'Problem of Time in Indian Thought'*, Calcutta Preview, No. 52, 1934.

Coomaraswamy, A.K.: *'Time and Eternity'*, Ascona, Artibus Asia Publishers, 1947.

Das, S.R. : *'Notion of Time in Indian Philosophy'*, Indian Historical Quarterly, No. 9, 1933.

Eliade, Maricea : *'Time and Eternity in Indian Thought', 'Man and Time'*, Eranos Year Book, London, Rontledge & Kagan Paul, 1958.

: *'Myths, Dreams and Mysteries'*, New York, 1960.

Fleet, J.T. : *'The Kaliyuga Era of B.C. 3102,* Journal of the Royal Asiatic Society, 1911.

Gupta, Rita Hindu : *'The Buddhist Doctrine of Momentariness and its Presuppositions'*, Dharmakīrti's Arguments in Hētu Journal of Indian Philosophy, 8, 1980, pp. 47-68.

Hirianna, M. : *'Advaitic Conception of Time'*, Poona Orientalis, 4, 1939.

: *'An Indian View of Present Time'*, Quarterly journal of the Mythic Society (Banglore) 14, 1924.

Inada, Kenneth. K. : *'Time and Temporality—A Buddhist Approach'*, Philosophy East and West, 24, 1974. pp. 171-79.

Jean Przyluski : *'From the Great Goddess to Kāla'*, Indian Historical Quarterly 14, 1938.

Jhaveri, Indukula : *'Concept of Kāla and Ākāśa in the Sāṅkhya Yōga System'*, Journal of Oriental Institute (Baroda), 5, 1955.

Kalupahana, David. J.: *'The Buddhist Conception of Time and Temporality'*, Philosophy East and West, 24, 1974, pp. 181-91.

Keith, A.B. : *'Time and Space (in Nyāya-Vaiśēṣika)'*, *'Indian Logic and Atomism: An Exposition of the Nyāya and Vaiśēṣika System'*, Delhi, Oriental Books Reprint Corpn., 1977.

Khare, R.S. : *'The Concept of Time and Time Reckoning among the Hindus, An Anthropological View Point'*, Eastern Anthropologist, 20, (1), 1967.

Koller, John, M. : *'On Buddhist View of Devouring Time'*, Philosophy East and West, 24, 1974, pp. 201-8.

Lancaster, Lewis. R. : *'Discussion of Time in Mahāyāna Texts'*, Philosophy East and West 24, 1974, pp. 209-14.

Mahadevan, T.M.P. : *'The Advaita View of Time'* (ref. Not found)

'Time and the Timeless', Madras, 1953.

Mc. Govern, W.M. : *'Cosmic Synthesis'*, A manual of Buddhist Philosophy, London, 1923, pp. 43-48.

Miya Moto, Shoson : *'Time and Eternity in Buddhism'*, Journal of Indian and Buddhist Studies, 7, No. 2, 1959, pp. 31-41.

Mukerjee, Satkari : *'The Buddhist Doctrine of Flux' (The Nature of Existence), 'The Buddhist Philosophy of Universal Lux'*, University of Calcutta, 1935, pp. 1-19.

Prasad Jwala : *'Discussions of the Buddhist Doctrine of Momentariness and Subjective Idealism in the Nyāya Sūtrās'*, Journal of the Royal Asiatic Society, 1930, pp. 31-39.

Puligandla, R. : *'Time and History in the Indian Tradition'*, Philosophy East and West, 24, 1974.

Sadananda Bhaduri : *'Time (in Nyāya-Vaśēṣika)'*, studies in Nyāya-Vaiśēṣika metaphysics, Poona, Bhandarkar Oriental Research Institute, 1975.

Sanatkumar Sen : *'Time in Sāṅkhya-Yōga'*, International Philosophical Quarterly, 8, 1968.

Sarma Arvind : *'The Notion of Cyclical Time in Hinduism'*, contributions to Asian Studies, 5, 1974.

Sasaki, Genjun. H : *'The Time Concept in Abhidharma'*, Proceedings of the 26th International Congress of Orientalists, New Delhi, 1964, pp. 471-80.

Satyavrat : *'The Concept of Time According to Bartṛhari'*, Annals of the Bhandarkar Oriental Research Institute, 39, 1959.

Staw, A.B. : *'Concepts of Time and Approaches to Analogical Reasoning in Historical*

Perspective', Bibl, American Antiquary, V. 58. 235-60. A.P. 93.

Stcherbatsky, Th. : *'Vasubandhu on the Fundamental Principle of the Sarvāstivāda School,' 'The Central Conception of Bhddhism and the meaning of the word 'Dharm'*, Royal Asiatic Society, London.

Von Wesendonk, O.N. : *'The Kālavāda and The Zervanite System'*, Journal of the Royal Asiatic Society of Great Britain and Ireland, 1931.

Williams, Paul. M. : *'Buddhadēva and Temporality'*, Journal of Indian Philosophy, 4, 1977, pp. 279-94.

Journals

Adyar Library Bulletin, Madras.

Annals of The Bhandarkar Oriental Resēarch Institute, Poona.

Annals of The Venkateswara Oriental Institute, Tirupati.

The *Bharati (A Telugu Miscellany),* Thambu Chetty Street, Madras.

Calcutta Oriental Journal, Calcutta.

Indian Antiquary, Bombay, Vols. 1-55.

Indian Culture, Calcutta.

Indian Historical Quarterly, Calcutta.

Indica, Heras Institute of Indian History and Culture, Bombay.

J.A.H.R.S. (Journal of the Andhra Historical Researhc Society).

Journal of Indian History, XLIII, Part-III, Dec., 1965.

Journal of the Asiatic Society of Bengal, Calcutta.

Journal of the Bombay Branch of Royal Asiatic Society, Bombay.

Journal of Oriental Research, Madras.

J.R.A.S. (Journal of the Royal Asiatic Society, etc.), London.

South Asiatic Quarterly-92, 139-55 World, S. Waldrop—*'The Contemporary Feature of Tomorrow'.*

INDEX

A

C

E

H

I

L

M

Q

R

T

W

Y

***